A TEXTBOOK OF PISCICULTURE AND AQUARIUM KEEPING

A TEXTBOOK OF PISCICULTURE AND AQUARIUM KEEPING

H.S. Jagtap
S.N. Mukherjee
V.K. Garad

2009
DAYA PUBLISHING HOUSE
Delhi - 110 035

ISBN10 81-7035-585-0
ISBN13 978-81-7035-585-4

Published by : **Daya Publishing House**
1123/74, Deva Ram Park
Tri Nagar, Delhi - 110 035
Phone: 27383999
Fax: (011) 23260116
e-mail : dayabooks@vsnl.com
website : www.dayabooks.com

Showroom : 4760-61/23, Ansari Road, Darya Ganj,
New Delhi - 110 002
Phone: 23245578, 23244987

Laser Typesetting : **Classic Computer Services**
Delhi - 110 035

Printed at : **Chawla Offset Printers**
Delhi - 110 052

PRINTED IN INDIA

Preface

Fisheries sector is not only contributed to the food basket of the country with quality protein component but also provides employment, rural development and export earnings in a big way. Aquarium fish culture is a potential money spinner not only to earn a living but also to make profits of exporting them. Ornamental fishes and plants are in greater demand due to their easy maintenance in houses, offices and commercial centers, the scope for their expansion is increasing day by day.

Considering the importance of food fishes and ornamental fishes various agricultural and non-agricultural universities include fishery subject in their syllabus. In this book the authors have tried to give comprehensive, to the point and up to date information.

The authors grateful to Prof. B.V. Jadhav [Department of Zoology, Dr. Babasaheb Ambedkar Marathwada University, Aurangabad (M.S.)] for their motivation in the production of this book.

We would like to thank all the distinguished authors and researchers whose work and contributions have formed a basis of some key material for this book.

The authors are also deeply indebted to his colleagues, relatives and friends who have helped in the preparation to this book.

The authors are extremely grateful to their educational institutes presidents, secretaries, directors, principals, librarians and family members for their sprit of co-operation during the writing of this book.

The interest taken by the publisher in this book is also highly appreciated.

Suggestions for the further improvement of the book shall be highly appreciated and would be incorporated in future edition of this book.

Authors

Contets

Part I
PISCICULTURE

Chapter 1

Introduction and Classification of Fish Culture

The knowledge of the occurrence of fish in India dates back to three millennium B.C. (Hora, 1956). Aquaculture is broadly defined as the farming of aquatic organisms such as prawn, mollusks, sea weeds, algae in addition to various fishes. "Pisciculture is the culture of fish by employing artificial means with scientific technique". Aquaculture improves the nutritional standards of the people and also it generates income and employment opportunities. Fish culture is in vogue for the past several centuries. The aquaculture in India has become an integral part of the rural development. In India and particularly in Andhra Pradesh, the fresh, brackish and marine water resources are vast. At present India is also one among the fish and prawn producing countries in the world scenario. Similarly inland fish production also is at reasonably high rate. Aquaculture technologies were rapidly developed in India in the fields of induced breeding for the production of large amounts of fish seed of good quality.

Fish occupies an important place in Indian mythology, history and tradition. One of the incarnations of God in India mythology was in the form of "Matsyavathara". Fish remains, indicative of their use as food by the people, were obtained in excavations at Mohenjodaro and Harappa of Indus Valley Civilisation (2500 B.C.–

1500 B.C.). The people of Egypt were probably the first in the world to culture fish (2500 B.C.). Fish farming is widespread in China since 2000 B.C. In India fish culture in reservoirs was mentioned in Arthasastra (between 321 to 300 B.C.) by Kautilya, indicate rendering fishes in reservoir poisonous in times of war (Hora and Pillay, 1962). King Somesvara of Chankya Dynasty in encyclopedia Manasoltrar complied in 1927 A.D. (Hora, 1951) described the method of flattering fish in the chapter as 'Matsya-Vinod". No further records are available till about the 19th century. Traditional fish culture existing in the eastern states of India particularly Bengal, Bihar and Orissa has gradually spread to the other regions by the end of the 19th century.

Fish enjoys a very special consideration and place in human civilization from times immoral. Fish receives greatest place in the national dish. Fisheries primarily started as capture fisheries indulging in commercial fishing in natural sources (seas: marine fisheries, rivers, lakes: freshwater fisheries). Fisheries and agricultural farming have evolved rather parallel in the history of human civilization. Interest in fish eating dates back to the dawn of history. It is believed that hunting of fish was not uncommon in prehistoric times. At the dwelling sites near a river or lake of cave dwellers of the late Old Stone Age (40000 BC) heaps of refuse of shellfish and sea fish have been found. The great distance by which these sites were separated from the sea points to some primitive way of fish preservation (like sun drying and smoke drying over the wood fire) practiced by these ancient dwellers to keep the food in edible condition so far upland. New Stone Age (10000 BC) has also given evidence of salmon smoking practices. Bronze Age (3500 BC) was the time when salting of fish started. Trading in dried fish, of course, was in vogue with ancient civilizations of Egypt, Mesopotamia and Indus valley.

The common goal fishery "the fish as food for mankind". Apart from the food–importance, there is one more importance it is the role it plays in the rural economy. Self employment in rural areas. Another great role of fishery industry lies in the alleviation of suffering of the people due to malnutrition. Fish is a chief source of animal protein. Egypt was probably the first in the world to culture fish (Marr *et al.*, 1966). But carp culture was wide spread in China in 2000 BC. Fish culture 1000 years of ago Khona-daughter of law of Varhamihir-recommended culture of vegetables in the embankments of fish ponds.

India is the 6th largest producer of fish in the world and 2nd in inland fishery. Presently India holds 3rd position in world with a total fish production. In India approximately 2500 different species found, out of that 930 from freshwater and 1570 from marine water.

Classification of Fish Culture

Fish culture practices may be classified in several ways, depending up on the several different aspects upon which emphasis is laid, singly or jointly, in a particular consideration. Three important considerations are given below:

A. Fish culture practices may be classified into three different kinds on economic and commercial consideration, depending upon the motive of farming: extensive, intensive and semi-intensive.

(*i*) Extensive Fish Culture

Is the least managed fish farming. Although the pond may be large, little care is taken with regard to its improvement. A modest yield, raised on the natural food, nothing more than the natural production, is obtained. Artificial food or fertilizers are usually not used.

(*ii*) Intensive Fish Culture

Is one in which an all-out attempt is made to achieve maximum production of fish from a minimum quantity of water. It is best managed form of fish farming, in which fish are fed on artificial food and the pond is improved by use of fertilizers. In such a case the small size of the pond notwithstanding, the production of fish is very high (over 6000 kg/ha/year) and of high quality fish. Although the cost of investments is also high, the return from the yield far exceeds it to ensure a great profit in the trade.

(*iii*) Semi-intensive Fish Culture

This is in between the above two kinds and is transitional stage. Full potential of the ponds is not utilized.

B. Fish culture practices may be classified into six basic categories, each with its own characteristics, depending on the nature of the water mass used on the manner of farming:

(*i*) Pond Fish Culture

In this case, the water-mass consists of natural or artificially erected freshwater ponds. It is the commonest of all methods.

(*ii*) Culture in Man-Made Reservoirs

Freshwater mass held back by the dam on reverse is used.

(*iii*) Paddy-Cum-Fish Culture

In this case, the paddy fields that are inundated with water are used for cultivation of fish.

(*iv*) Brackish water Fish-Culture

Open or embanked estuaries with waters of varying salinities, afford low lying, shallow fields under water for fish rising.

(*v*) Coastal Fish Culture (Mariculture)

Marine species of fish are cultivated in marine coastal waters alone.

(*vi*) Salt Pan Culture

In off-seasons of salt manufacture, the salt pans are utilized for fish culture.

C. Classification of fish culture practices based on special operational techniques is as follows:

(*i*) Cage Culture

Here the fish to be cultured are held in a section of water of a following river by being imprisoned in a cage of metal or bamboo structure, supporting wire, nylon or split-bamboo mesh mats. It is rather recent practice but with great promise. Generally, carnivorous species are cultured in cages.

(*ii*) Pen Culture

This is simply an enclosure, often along banks and shores, of some device which utilizes the bank or the shore line as one of its sides. Fishes are then raised in this enclosure.

D. Classification based on the number of species cultivated at a time.

(*i*) Monoculture

Only one species of fish is cultured in a pond.

(*ii*) Monosex Culture

In this case, only one sex, either the male or the female, of a single species is cultivated in a water body. Tilapia is generally used for mono sex culture.

(*iii*) Polyculture

In this case, more than one species is raised simultaneously in the same water body.

E. Fish culture practices may also be classified on considerations of climate factors:

(*i*) Warm Water Culture

This concerns water masses of the tropics and plains.

(*ii*) Cold-Water Culture

This involves culture practices conducted in springs, pools or lakes of hills and high altitudes.

Chapter 2

Morphology and Anatomy of Any Cultivable Fish (*Catla*, *Rohu*, *Mrigal*)

Labeo rohita

Morphology

Shape, Size and Color

Body is spindle-shaped. Color is greyish or blackish on back and silvery white or pale on the two sides and belly. A full grown individual measures 1 meter in length and 20 to 25 kg in weight. The body is divisible into head, trunk and tail.

Head

Head extends from tip of snout to hind edge of operculum. Snout is depressed, short and obtuse. Mouth is sub-terminal. It is a large transverse aperture bounded by thick and fleshy lips. Usually only two small, thread-like sensory maxillary barbels are present at the corners of mouth, the rostral barbels being generally absent.

Teeth are lacking on jaws. Snout bears dorsally a pair of small nostrils. The lateral eyes on head are without eyelids. Behind the eye on either side is a large movable bony gill cover or operculum with free posterior and ventral margins. Beneath each operculum lie four comb-like gills in a brachial chamber.

Trunk

It is the thick middle part of body. It is higher than wide and oval in cross section. On either side of trunk and tail is a lateral line.

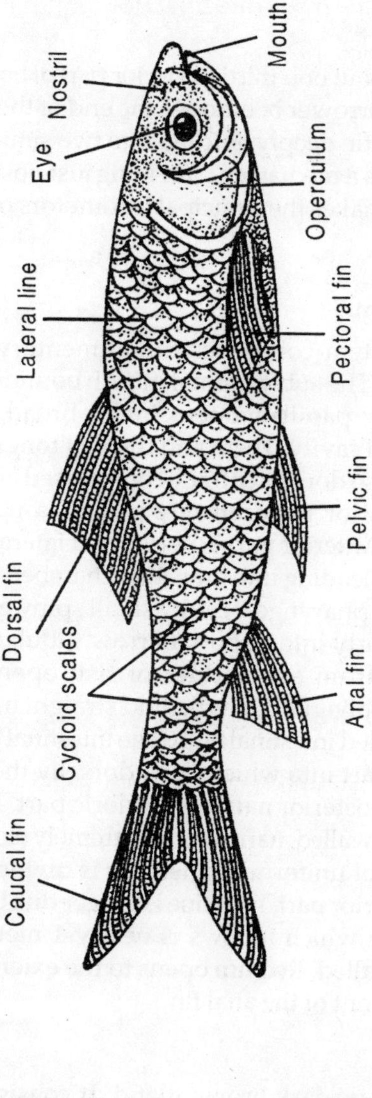

Figure 2.1: *Labeo rohita*

Fins are well developed and supported by bony fin rays. On the back of the middle of trunk is a single large somewhat rhomboidal dorsal fin. Just behind operculum is a pair of large ventro-lateral pectoral fins, followed by a pair of smaller ventral pelvic fins. Mid-ventrally at the posterior end of trunk lie in a series three small apertures: anterior anus, middle genital and posterior urinary.

Tail

It comprises about one-third posterior part of body. It is laterally compressed and narrower behind. At the end of the tail is a median homocercal caudal fin deeply notched into two similar lobes. On the ventral side of tail is a median anal fin lying just posterior to urinary aperture. The tail makes the principal locomotors organ.

Anatomy

Digestive system

Digestive system comprises the alimentary canal and the associated glands. The sub-terminal mouth bounded by fleshy lips beset with sensory papillae, opens into a broad, dorsoventrally compressed buccal cavity. Teeth and distinct tongue are lacking in rohu. Pharynx is also dorso-ventrally compressed and differentiated into a broad anterior respiratory part and a narrow posterior masticatory part. Anterior part is perforated laterally by four pairs of internal gill slits leading into branchial chambers. From branchial arches project into pharyngeal cavity small spiny gill slits. Pharynx leads into posteriorly into a short, narrow tubular oesophagus. A pneumatic duct from air bladder of fish opens dorsally into oesophagus. Oesophagus opens behind straight into an elongated, swollen, thick-walled intestinal bulb. The intestinal bulb has anterior broader cardiac part into which open dorsally the pancreatic and bile ducts, and a posterior narrower pyloric part. Intestinal bulb is followed by a thin-walled, narrow and extremely elongated intestine. It is a coiled tube of uniform diameter. It is divided into two parts anterior and posterior part. Intestine is longer due to its herbivorous habit. The rectum which follows is nearly 1 meter long, slightly wider and thin-walled. Rectum opens to the exterior by anus mid-ventrally just in front of the anal fin.

Liver

Liver is a large, dark brown gland. It consists of a narrower right lobe and a broader left lobe interconnected anteriorly, in the

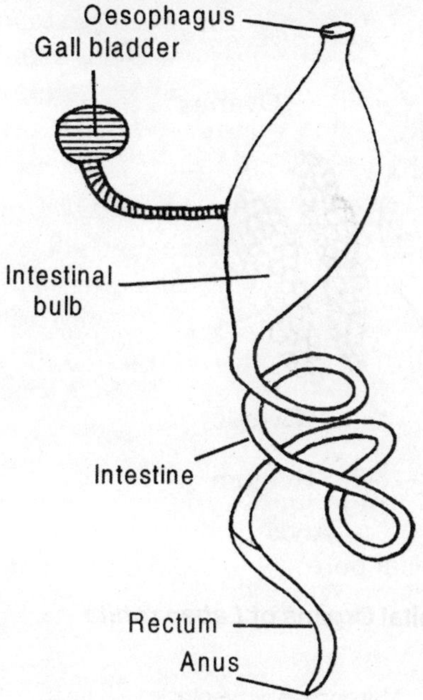

Figure 2.2: Alimentary Canal of *Labeo rohita*

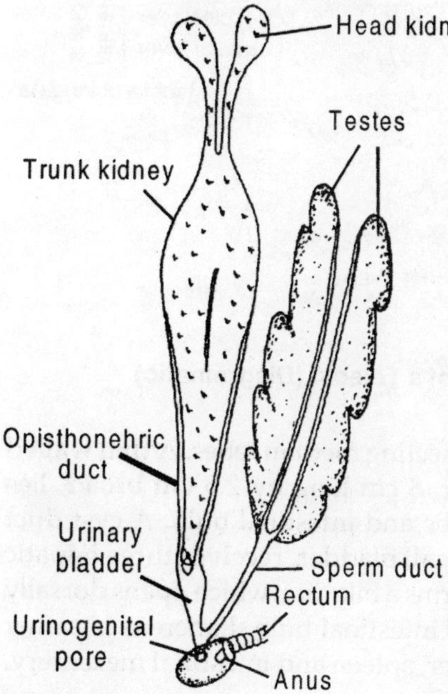

Figure 2.3: Male Urinogenital Organs of *Labeo rohita*

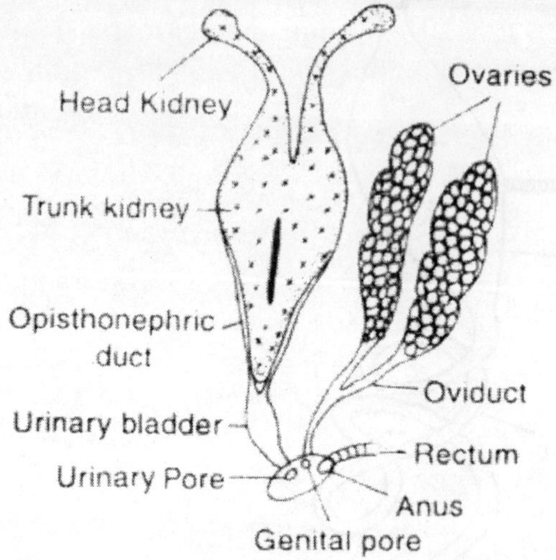

Figure 2.4: Female Urinogenital Organs of *Labeo rohita*

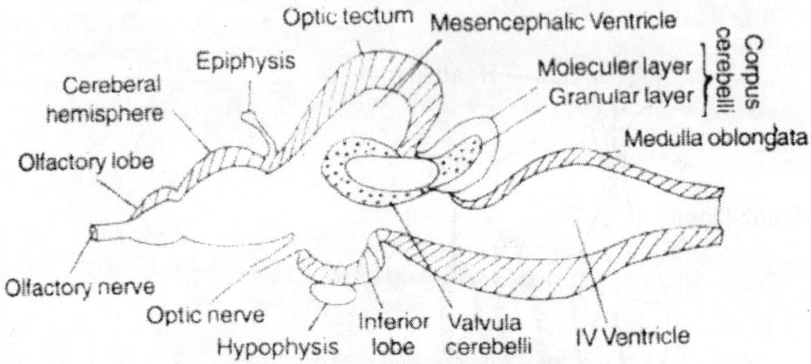

Figure 2.5: L.S. Brain of a Teleost (Diagramatic)

middle and posteriorly, by connecting median lobes. A thin walled elongated sac like gall bladder, 8 cm long by 2.5 cm broad, lies dorsally between right liver lobe and intestinal bulb. A cyst duct arises from the anterior end of gall bladder, receives three hepatic ducts from the liver lobes and forms a bile duct which opens dorsally into the roof of cardiac part of intestinal bulb. Pancreas is rather diffused, found scattered in liver, spleen and intestinal mesentery.

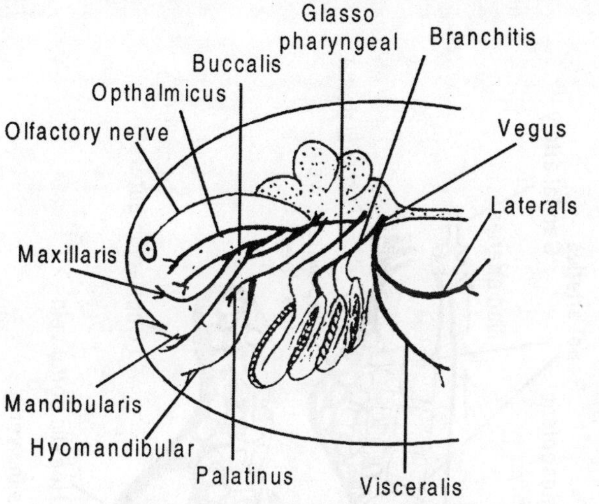

Figure 2.6: Cranial Nvers of *C. mrigala*

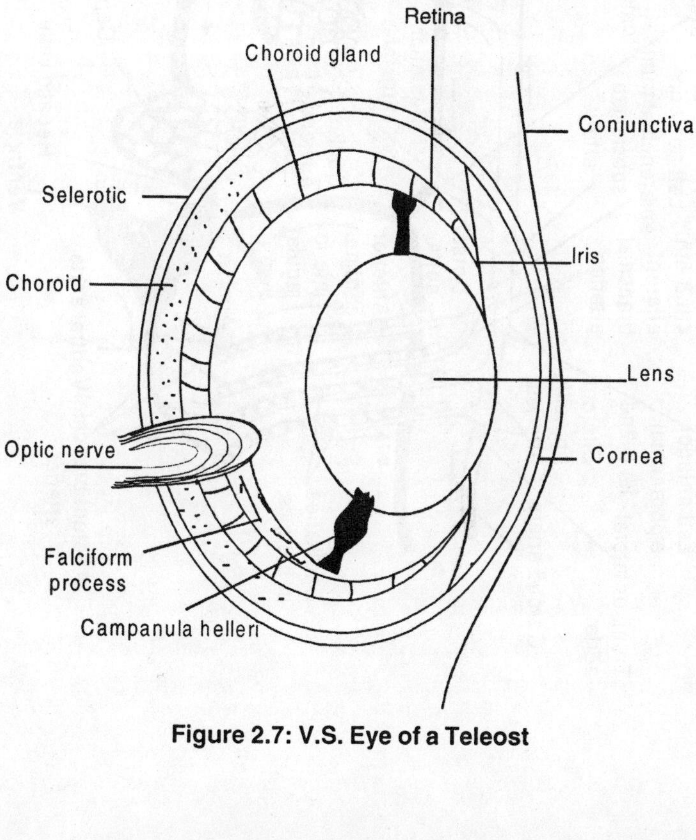

Figure 2.7: V.S. Eye of a Teleost

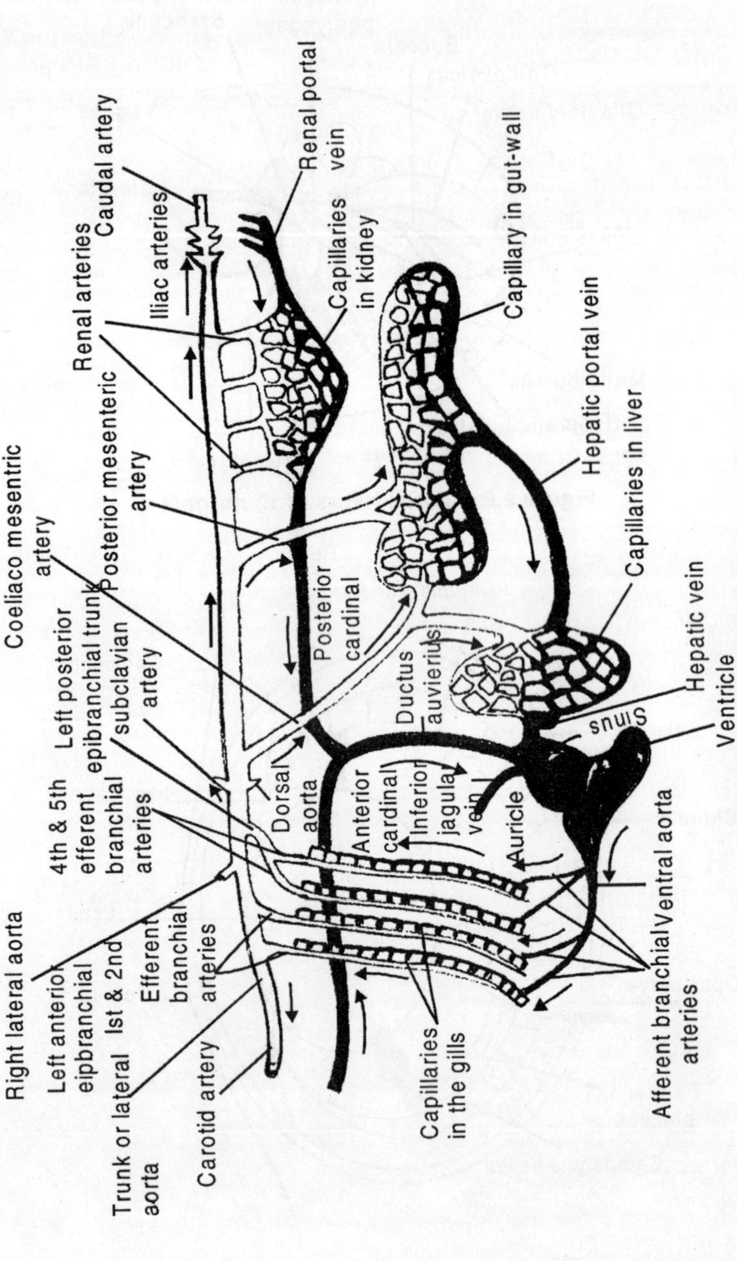

Figure 2.8: Circulatory System of *Labeo rohita*

A pancreatic duct, which is enclosed in a common sheath with bile duct, opens separately in the intestinal bulb.

Air Bladder

Air bladder is also called the swim bladder. It lies in the body cavity, but outside coelom, above the intestinal bulb and ventral to the vertebral column. It is an elongated white, thin walled sac filled with a gas. It is divided by a transverse constriction into a small anterior chamber and a large posterior chamber. The air bladder is connected to the roof of oesophagus by a slender pneumatic duct.

Gill

Four pairs of filiform gills supported by the first four pairs of gill arches. Four gills are present on either side of pharynx in a

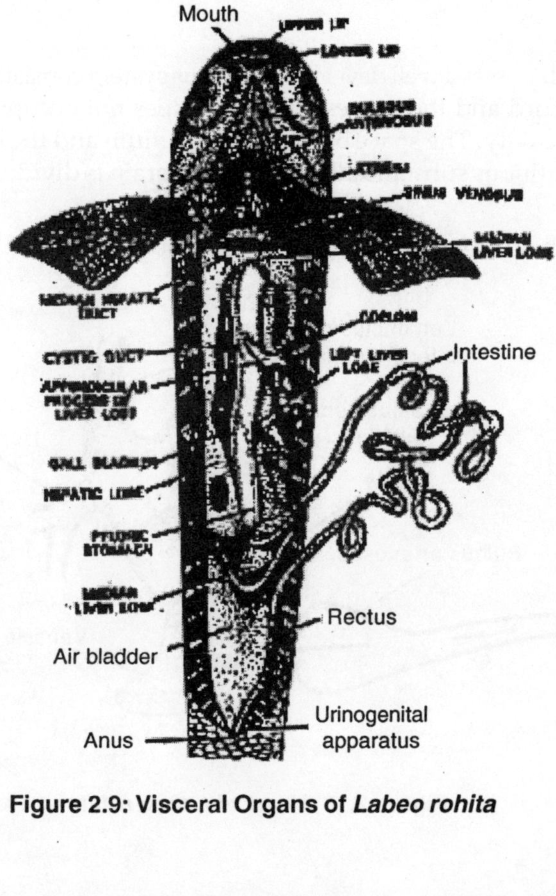

Figure 2.9: Visceral Organs of *Labeo rohita*

common gill chamber. Each gill chamber is covered externally by a skin flap, the operculum, supported by bony plates. The chamber opens to outside by a large crescentric branchial or gill aperture behind the operculum and in front of the pectoral fin. Each gill consists of two rows (hemibranchs) of slender gill filaments.

Heart

The 2-chambered heart lies in an anterior portion of coelom, the pericardium, beneath pharynx. Sinus venosus is large and spongy and bears a pair of lateral appendages. It opens into thin-walled auricle, thence into thick-walled ventricle, all having walls to prevent reverse flow of blood. An enlargement of ventral aorta, the elastic bulbus arteriosus is present. Short ventral distributes blood to gill capillaries by 4 pairs of afferent branchial arteries. 4 pairs of efferent branchial arteries into dorsal aorta.

Brain

Fish posses a well developed nervous system consisting of brain, spinal cord and the nerves. The brain does not occupy the entire cranial cavity. The space between the cranium and the brain is full of a gelatinous substance. The teleostean brain is divided into three regions:

(*a*) Prosencephalon (fore brain)

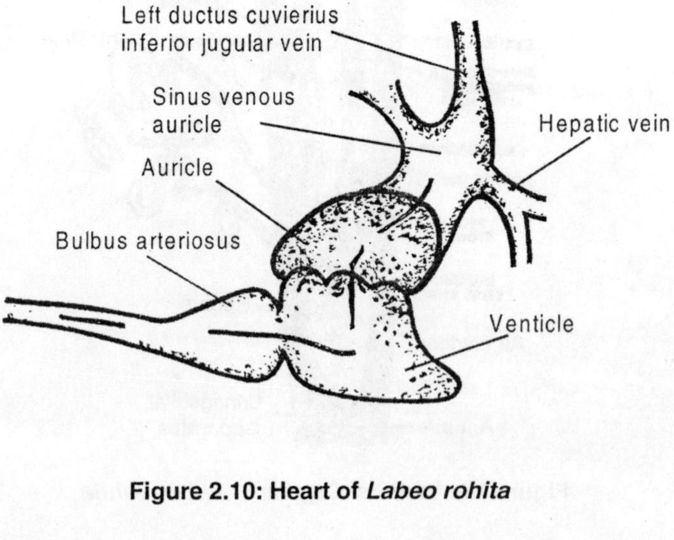

Figure 2.10: Heart of *Labeo rohita*

(*b*) Mesencephalon (mid brain), and

(*c*) Metencephalon (hind brain).

Fore Brain

It consists of telencephalon and the diencephalons. Of these, the telencephalon includes a pair of solid olfactory lobes and two large cerebral hemispheres. A pair of short or long olfactory nerves arise from the olfactory lobes and run forward up to the olfactory rosette. A small swelling at the base of the olfactory rosette may be present and is called the olfactory bulb.

Mid Brain

The mid brain consists of a dorsally situated optic tectum and a ventral tegmentum. The optic tectum forms two optic lobes, and is made up of at least five zones. The optic tectum projects into the optocoel, and forms a pair of torus longitudinalis, which is a characteristic feature of all the teleosts. The torus connects the tectum with the posterior commissure. The cavity of the optic lobes becomes obliterated due to the presence of valvula cerebelli, which consists of a granular and a molecular layer.

Hind Brain

The hind brain consists of the cerebellum and the medulla oblongata. The cerebellum is formed as a dorsal outgrowth from the hind brain, but only a small part of it can be seen from outside, as corpus cerebelli. The anterior part of the cerebellum enters into the optocoel and forms the valvula cerebelli. The corpus cerebelli consists of an outer molecular layer and an inner granular layer, which are reversed in position in the valvula cerebelli.

The spinal cord extends for the entire length of the body from the head up to the end of tail. Two kinds of nerves are present in fishes, the spinal nerves and the cranial nerves. The spinal nerves are arise in pairs from the spinal cord and are metamerically arranged. Cranial nerves originate from the brain and ten pairs are present in teleosts.

Urinogenital System

Sexes are separate and excretory and reproductive organs are quite in dependent. Two elongated and brownish mesonephric or opisthonephric kidneys lie just beneath the vertebral column in abdominal cavity. Their free anterior ends have knob like head kidneys. Their broad middle and narrow posterior parts are united and form the renal part containing uriniferous tubules. Each kidney in a slender, tubular ureter. Both the ureters unite into a common urinary duct with a thin walled urinary bladder. Bladder leads into midventral urinary aperture situated behind the anus.

One pair of elongated gonads, testes in male and ovaries in female, lie posteriorly in the abdominal cavity beneath the air sac. They are suspended from the body wall by folds of peritoneum, called mesorchia in male and mesovaria in female. The mature gonads become exceptionally enlarged during breeding season. A pair of gonoducts are formed by peritoneum, called sperm ducts in male and oviducts in female. Both ducts join posteriorly to open by a single genital aperture lying mid-ventrally between anus and the urinary aperture.

Chapter 3
Fish Breeding

Both male and female gonads undergo marked cyclic morphological and histological changes before reaching full maturity and becoming ripe. This is called maturation of gonads. Most of the fishes exhibit seasonal cycle in the production of gametes. Gametes expelled out from the body into the surrounding water is called spawning resulting in fertilization. Fish spawns during a specified period which the gonads attain full maturity and spawning takes place in the population is called breeding season of the species.

The important cultured fishes according to their breeding habits may be grouped into three distinct categories:

1. Fishes which breed in ponds, *e.g. Cyprinus carpio, Tilapia mossambica, Osphronemus goramy,* Channa species, *Etroplus suratensi,* etc.
2. Fishes which do not ordinarily spawn in ponds but breed in flooded rivers and adjoining area, *e.g. Catla catla, Labeo rohita, Cirrhinus mrigala, Labeo calbasu, Hypopthalmichthys molitrix, Ctenopharyngodon idella,* etc., and
3. Fishes which mainly inhabit the sea or brackish waters and which neither spawn in ponds nor could be induced so far to spawn success fully therein, etc.

The breeding requirements of important cultured species are complex and optimum conditions. A dependable source of quality fish seed is a fundamental prerequisite for large-scale development of fish culture. Fish seed consists of the eggs, hatchlings, fry and fingerlings.

Indian major carps are known to breed naturally in rivers and certain reservoirs and in artificially constructed bundh-type tanks where water flowing conditions are stimulated during the spawning season. Fish seed is generally collected from breeding grounds in rivers but such a collection is a mixture of economic and uneconomic species. It is also mixed with eggs of predatory species and it is difficult to separate them, at the early stage, rear them to an identifiable size when they could be separated. The rearing of number of unwanted fish involves wastage of valuable nursery space and expenditure.

Hence several methods have been developed to obtain fish seed by artificial methods which include artificial fertilization and induced breeding.

Artificial Fertilization by Stripping

Eggs of certain species of carps can be artificially fertilized by stripping. There are two methods: a) Wet, and b) Dry.

Wet Method

In wet method, the milt and eggs are striped directly into the water. Wet method is preferred if the eggs are sticky.

Dry Method

In dry method water is added after the eggs and sperms have been thoroughly mixed. This method generally used in non-adhesive eggs.

The female is caught first and dried with the help of a towel. The fish is held with the left hand, its belly turned upwards and stripping is done with the thumb of the right hand. The eggs are forced out by gently massing the abdomen, starting from the hinder end first. After the end of stripping all the eggs have passed out into the container. Then a male is caught, dried with a towel and kept with belly downwards. The milt is pressed out by stripping into the same container having eggs and thoroughly mixed with the help of a feathers. Water is added and eggs mixed again by stirring. After sometime, the fertilized eggs are transferred to hatcheries or hapas for hatching.

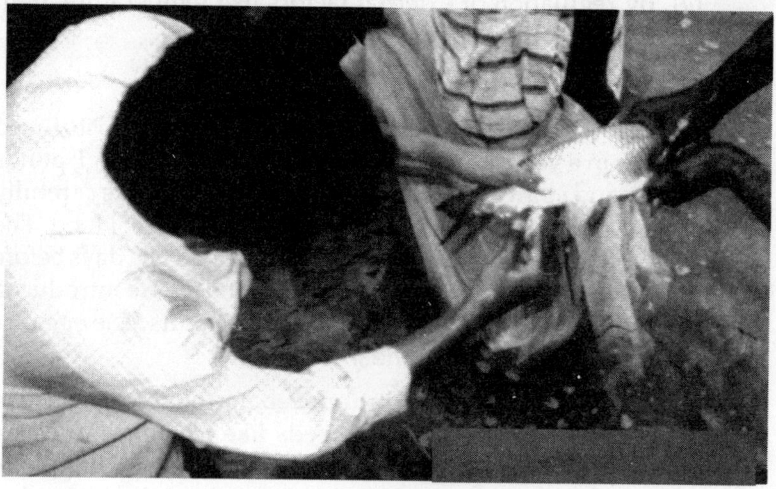

Figure 3.1: Collection of Milt

Methods of Induced Breeding of Fishes

According to their breeding habits and breeding requirements, the cultivated species may be divided into two main groups.

1. Pond breeding fishes and species that easily breed in ponds
2. Species that do not easily breed in ponds, and special methods are required to induce them to breed.

1. Method Adopted for Pond Breeding Fishes

Pond breeding fishes also need some inducement. Wild spawning can take place in ponds, but the survival of young ones is poor. Hence, pond breeding species are segregated sex-wise to prevent wild spawning in ponds, and are then induced to breed by one of the following methods:

(*a*) A few selected brood fishes are introduced in special breeding ponds containing freshly introduced well oxygenated water, where spawning takes place.

(*b*) Spawning is induced by simple replacement of old pond water with fresh rain water.

(*c*) By providing suitable material for attachment of eggs.

(*d*) By regulation of external factors like temperature, light and population density.

(*e*) By injecting hormones.

The common carp *Cyprinus carpio* is a pond breeding fish. Breed several times in a year. Separate ponds for brood fishes and, ponds for hatching and spawning are constructed. Breeders are carefully selected, segregated sex-wise, and are fed with artificial food. The spawning pond has a hard bottom and is dried for a few days before filling it with oxygen rich water. Selected brooders are introduced into the pond. Spawning takes place within 16 hours. The eggs are then transferred to hatching ponds.

In India, selected fishes are introduced into cloth hapa fixed in ponds and supplied with aquatic weeds like *Hydrilla* and *Najas*. Fishes breed within 10-12 hours after introduction and the eggs are transferred to hatching hapas. *Cyprinus carpio* can also be bred in cement cisterns filled with tap water.

Fishes like *Clarias batrachus*, *Heteropneustes fossilis*, and *Anabas testudineus*, can be induced to breed in ponds and cement cisterns by mere replacement of old water by fresh rain water.

2. Methods Adopted for Difficult to Spawn Fishes

Fishes belonging to the second group do not ordinarily breed in ponds. These includes cultivated species like the Indian major carps, Chinese carps, a few commercial catfishes and brackish water species like the mullet, milk fish, etc. Majority of these fishes have been induced to breed by developing various techniques. There are two methods:

(*a*) Breeding in special type of ponds called "Bundhs".

(*b*) Breeding by means of hormones

Bundh Breeding

Bundhs are special type of perennial and seasonal tanks or special tanks where riverine conditions are simulated during monsoon months. Bundhs are of two types, *viz.*, perennial or wet bundh, and seasonal or dry bundh.

(*a*) Perennial or Wet Bundh

It is a perennial pond situated in the slope of a vast catchments area, with embankment on three sides. The main pond retains water throughout the year, but its shallow marginal areas dry up during the summer months. The bundh has an inlet towards the high catchment area, and an outlet at the opposite lower side. The bundh usually gets flooded with water from the upland area after heavy showers. The shallow area of the bundh called 'moan' gets inundated and excess of water flows out. The outlet is protected by bamboo fencing. The breeders which are either grown the perennial pond or released from other ponds, get stimulated by the flow of silt laden, well oxygenated rain water. During summer the deeper portion of the pond retains water containing breeders. The remaining portion is dry and is used for agriculture. The wet bundhs are comparatively much bigger in size than the dry bundhs. In wet bundhs the brood stock may be maintained through out the year or replenish prior to monsoon.

The flow of the water through the outlet can be controlled by blocking with straw and mud.

(*b*) Dry Bundh

A dry bundh is a seasonal shallow pond enclosed by an earthen wall on three sides. During the monsoon season, rain water rushes from the vast catchment area and accumulates in the pond. Breeders from nearby ponds are introduced in the shallow ponds in the ratio of 1 female and 2 males.

A large number of bundh type of tanks are found in W. Bengal, Bihar and Orissa.

Stocking density 3000 to 3500 kg/ha.

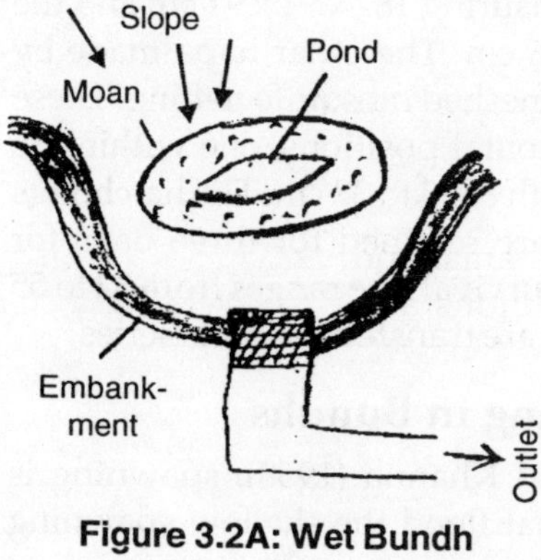

Figure 3.2A: Wet Bundh

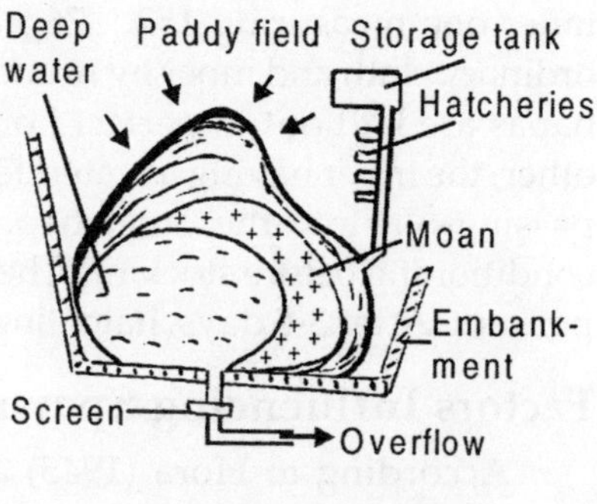

Figure 3.2B: Dry Bundh

Spawning

Spawning in both wet and dry bundhs usually occur after a heavy shower, when large quantities of water rushes into the bundh. It has been observed that the fishes migrate to shallow water, male and female exhibit sex play for a short time during which their bodies are twisted round each other. The pressure on the abdomen results in the exudation of ova and milt. Eggs are laid at intervals during which the pair keeps moving. After spawning the spent fish move to the deeper water.

Dry bundhs are considered to be one of the most reliable means for mass breeding of grass carp and silver carp. Spawning occur at night and during the bright sun in the forenoon. Mrigal and rohu occupies marginal area and breed in the forenoon. Catla may occupy deeper water due to its big size and breed from noon till evening. The fertilized eggs are left behind by the parents. The temperature of water at spawning grounds has been reported to range from 24 to 32°C in various environment.

Collection of Eggs and Spawn

Soon after the spawning process is over, the eggs are collected from the bundh with the help of nets made of mosquito netting cloth (Gamaha). The eggs collected are released to small hatching pits or in double walled hatching hapas or in cement hatcheries. The hatching pits of the size of 448 x 244 x 46 cm are excavated on the bank with arrangement for the supply of water. Each pit of this size may contain about 9,00000 to 2,20,0000 eggs of the size of 4′ x 3′ x 3′ for hatching. A double walled hapa, which is fixed in the bundh itself, consists of an outer hapa measuring 182 x 91 x 91 cm and the inner one measuring 152 x 76 x 46 cm. The outer hapa made by ordinary cloth and inner by round meshed mosquito netting. These hapas are laid out in perfect horizontal positions, one within the other, the inner having a water depth of 23 to 32 cm. The hatchlings passing out into the outer hapa are retained for three days for conditioning before stocking. The survival rate ranges from 32 to 55 per cent. After 3-5 days, hatchlings are transferred to nurseries.

Factors Influencing Spawning in Bundhs

According to Hora (1945) and Khanna (1958), spawning is stimulated by heavy monsoons that flood the shallow spawning areas. It appears that availability of shallow spawning grounds inundated with fresh rain water is an important factor in stimulating the breeders. Strong water current on the spawning grounds favors breeding. The favorable temperature of water for spawning has been found to vary from 24-32°C. Generally cloudy days followed by thunderstorm and rain, is regarded to influence spawning. Other factors like pH, alkalinity, high oxygen content etc., may be secondary importance, and are associated with floods.

Spawn collected from the rivers and wet bundhs, besides comprising a mixture of numerous species, many of which are predaceous. There is need to develop a special induce breeding technique so as to procure pure fish seed.

(b) Breeding by Means of Hormones

Induced Breeding by Hypophysation

The Indian major carps, catla, rohu, mrigal and Chinese carps breed in the rivers during monsoon season. The seed collected from rivers consist not only of preferred species and also may include uneconomic and even predatory forms.

Pituitary gland play an important role in the maturation of the gonads and spawning in fishes. Induced breeding is a technique whereby ripe fish breeders are stimulated by pituitary hormone injection to breed in captivity is called hypophysation.

The Brazilians were the first to succeed in inducing spawning by injecting fish with pituitary extract. The Indian and Chinese carps are induced to breed by the administration of fish pituitary hormones and the method has become very popular for obtaining commercial production of seed of the cultivated species.

The concept of the application of pituitary injections for successful spawning of fish is due to Houssay (1930) of Argentina. Brazil was the first country to develop a technique for hypophysation. Von Ihering and his co-workers conducted experiments with various hormone injections on the lines of Houssay (1930) and achieved success in 1934. Since then, Brazilian fish culturists have been employing this technique to obtain seed. Major breakthrough in hypophysation in India came to the limelight in 1957, when success was achieved in breeding of Indian major carps, by Chaudhari and Alikunhi with carp pituitary extracts. Since then, the employment of this technique has spread widely and now forms a regular part of piscicultural programmes in many parts of India. Chinese carps have also been successfully bred in 1962 by employing similar technique. In India, Khan (1938) was the first to successfully induce *Cirrhina mrigala* to spawn by injecting mammalian pituitary hormone, and Chaudhari (1955) succeeded in inducing *Esomus danricus* to spawn by intra-peritoneal injection of pituitary gland from catla. Ramaswamy and Sundararaj (1956) succeeded in inducing spawning in *Heteropneustes fossilis* and *Clarias batrachus*. Later, several carps *Labeo rohita, Labeo bata, Cirrhina mrigala, Cirrihina reba* etc., were successfully bred by injecting pituitary extract (Chaudhari and Alikunhi, 1957). Now this technique is very popular in obtaining fish seed for commercial purposes. The hypophysation technique as follows.

Location of Pituitary Gland

The pituitary is situated on the ventral side of the brain just behind the optic chiasma and is attached to the infundibulum by means of a stalk.

Stage of Maturity of Donor Fish

Pituitary glands collected from fully ripe gravid fishes, as the gland is most potent at the time of breeding or just before spawning. In India, the most suitable time for collection of pituitary glands of major carps is during the months of May, June and July. Common carp (*Cyprinus carpio*) is an excellent donor, because the mature specimens of this species are available throughout the year all over India.

Methods of Gland Collection

The pituitary gland may be collected by the following two types:

(*i*) By dissecting and removing a portion of the scalp, or

(*ii*) Through the foramen magnum

(i) By Dissecting and Removing a Portion of the Scalp

A portion of the brain-case or scalp just above the brain is cut from the top with an oblique cut stroke by a hand-saw or sharp butcher's knife. A big bone-cutter (220-25 cm size) may be used in dissection of smaller fishes. After removing the portion of the scalp the gray matter and the fatty substances over the brain are seen. The gray matter and fatty substances over the brain are next removed gently clearing them with a piece of cotton held by forceps and the dorsal side of the brain is seen clearly. The olfactory and optic nerves at the anterior end of the brain are then detached by means of tweezers and the entire brain is then lifted up and laid back. The pituitary gland is located ventrally just posterior to the optic nerve crossing (optic chiasma). In some fishes where the gland is found attached to the brain. In carps pituitary gland is separated by a thin membrane from the brain. When the brain is lifted, the gland is left behind on the floor of the brain box in the concavity (Sella turcica). The gland is then carefully picked up by a pair of fine tweezers and kept in a cavity block or a petridish in absolute alcohol, under cover. Clearing the muscles.

(ii) Through the Foramen Magnum

The pituitary gland can be collected by this method from the cut fish-heads only. If the cut fish heads have more of flesh, the extra muscles with the portion of the condyles and the part of the first vertebra which are left with the head are cut out by means of a bone-cutter and removed, and the foramen is clearly seen with the gray matter and fatty substances in it. The brain lies on the ventral side of the foramen. The gray matter and the fatty substances over the brain are then removed by inserting the blunt end of the forceps into the foramen and dragging the substance out posteriorly with care, so that the brain is not detached or damaged. The foramen is then cleaned with a piece of cotton, to expose the dorsal region of the brain.

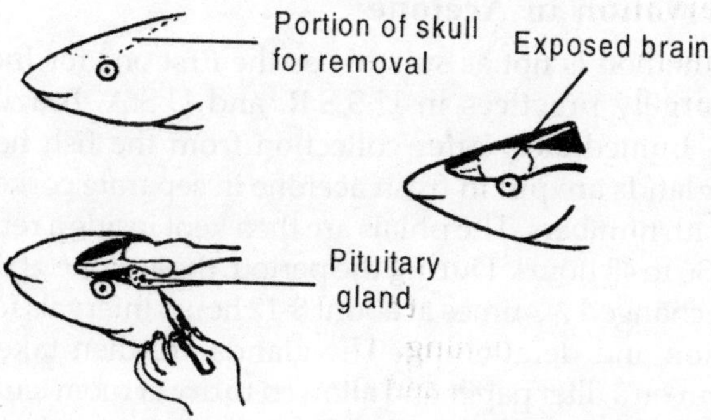

Figure 3.3: Removal of Pituary Gland from Fish

The anterior end of brain is first detached by means of a pair of fine forceps inserted into the foramen and then the entire brain is lifted up carefully and taken out of the whole. The pituitary gland is then seen under the membrane. The membrane is detached posteriorly by a pair of forceps and lifted up, and the pituitary gland is exposed. The gland is then picked up carefully by means of tweezers.

Preservation and Storage of Pituitary Glands

There are three principal methods of preservation of fish pituitary glands:

(*i*) Preservation in absolute alcohol

(*ii*) Preservation in acetone

(*iii*) Preservation by immediate freezing.

(*i*) Preservation in Absolute Alcohol

The pituitary glands are put in absolute alcohol immediately after collection. Each gland is kept in a separate phial marked serially to facilitate identification. After 24 hours the gland is washed with absolute alcohol and kept again in the phial with fresh absolute alcohol and stored in cool, shady place at room temperature. The phials must be air-tight especially when kept at room temperature and the alcohol should be changed from time to time. In order to prevent moisture from getting inside a desicator containing some anhydrous calcium chloride. The gland retains its potency in this method for 2-5 years.

(*ii*) Preservation in Acetone

This method is not as suitable as the first one for India. This method largely practices in U.S.S.R. and U.S.A. Brazilian fish culturists. Immediately after collection from the fish heads, the pituitary glands are put in fresh acetone in separate corked phials marked with numbers. The phials are then kept inside a refrigerator (50°F) for 36 to 48 hours. During the period, the acetone of the phials should be changed 2-3 times at about 8-12 hours intervals for proper dehydration and defattening. The glands are then taken out of acetone, put on a filter paper and allowed to dry at room temperature for one hour. The weight of the dried gland is then taken individually or in a group in a chemical balance and the glands are transferred to separate sterile phials marked with number. The phials with the glands are then stored till required inside a refrigerator at 50°F. The gland retains its potency for 6 months.

(*iii*) Gland Preservation by Immediate Freezing

Fresh glands are frozen immediately after collection from the fish-heads and kept in frozen storage in a refrigerator or in insulated cans containing dry ice.

Preparation of Pituitary Extract

Just before injection, previously weighed glands are taken out and homogenized in distilled water or 0.3 per cent saline in mortar and pestle. The suspension is then centrifuged and supernatant fluid is taken in a syringe for injection. Extract is also prepared with distilied water and glycerine in ratio 1:2.

Collections of Breeders for Injection

At the time of injection the breeders are netted out with a drag net and the suitable ones are segregated from the catch and collected by means of hands nets. The breeders are collected from the drag net one by one in the hand-net, and then the weights of the breeders along with the hand-nets are taken by means of spring balance. The actual weight of the breeders is calculated by deducting the known weight of the hand-net from the total weight of the fish along with the net.

Selection and Identification of Brooders

For success in induced spawning, good brooders are selected well in advance and kept in breeding ponds and fed with artificial food. Proper identification of sex and stage of maturity is an important requisite for good results. At commencement of the rainy season, male and female are segregated in separate ponds.

Male: Oozes milt at slight pressure on the abdomen and roughness on the dorsal surface of pectoral fins.

Female: Soft bulging abdomen and swollen vent.

It is preferably to choose healthy breeders in the age group 2- 4 years with their weights ranging from 1 kg to 5 kg.

Figure 3.4: Selection of Broodstock

Injection of Breeders

The required suspension is taken in a hypodermic syringe for injection into the breeders. The breeders in the hand-nets are then brought one by one for injection from the enclosure of the net and placed on a small field table provided with a cushion. Intramuscularly on the back or base of the caudal fin or near the

Figure 3.5: Intramuscular Injection

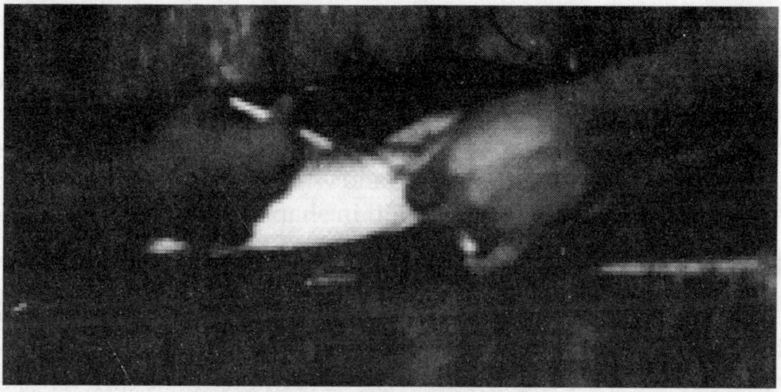

Figure 3.6: Intraperitonial Injection

shoulder region. For intramuscular injection, the needle is inserted under a scale, first parallel to the body of the fish and then pierced into the muscle at an angle. Some times intra-peritoneal injection is given at the base of pectoral fin. Intra-muscular injections are found to be more commonly given.

The most convenient hypodermic syringe used for the purpose is a 2 cc one having graduations to 0.1 cc division (needle No. 22 for 1 to 3 kg fish; No. 19 for large ones).

Dosage of Pituitary Extract

The breeders are selected carefully and weighed so that the quantity of the pituitary gland to be injected can be calculated. One set usually consists of one female and two males. Dose for carp species average female is generally given a stimulating dose of 2-3 mg of pituitary/kg body weight of the fish. After an interval of 6 hours the female is given a second higher dose of 5-8 mg/kg body weight and the males are given a low dose of 2-3 mg/kg body weight. Both the injected males and one female are now introduced into a cloth hapa, fixed in a bamboo poles in a pond. A breeding hapa is a rectangular box shaped container made of mosquito net cloth, and are closed on all sides to prevent the breeders from escaping.

Response of Injection

After about 2 hours of the second injection the breeders start swimming actively, excited and restless and try to jump out. The more active one chases the other, pushing it with the snout. The male does not allow the female to rest. Spawning usually occurs within 3-6 hours after the second injection. The eggs are examined to evaluate the percentage of fertilization. In carps fertilized eggs are transparent and look like pearls due to swelling up of eggs, unfertilized eggs become opaque and whitish, and non-floating and non-adhesive in nature and round in shape.

Estimation of Egg Quantity

The eggs are collected from breeding ground by means of cup or tray or beaker and transferred to the buckets. The breeders are also removed from the hapa, their weights are noted. The difference in weights reveals approximately the number of eggs laid. The eggs are kept in a rectangular piece of close meshed mosquito net and allow the water to drain off. The eggs are measured in a beaker, mug or cup of known volume and transferred to hatcheries. Thus estimation of total quantity is made from total volume of the eggs measured.

Treatment to Injured Fish

During breeding activity fishes get injured such fish give a bath of 1 to 2 ppm potassium permanganate before releasing into stocking pond.

Synthetic Hormones in Induced Breeding of Fishes

Induced breeding of fishes with the pituitary injection is becoming very popular, and provides a reliable method of obtaining pure seed for pisciculture.

Now various ovulating agents reported for induced breeding, that replace the hypophysation technique. Mammalian hormones human chorionic gonadotropins have provided good results which help for maturation of gonads in fish. Its activity like LH.

Ovaprim

It is a synthetic induce breeding hormone developed by Dr. R. E. Peter's research group of Canada and Dr. H. R. Lin's research group of China. The scientists named this new technique as a "LINPE METHOD", in 1988. This synthetic hormone is named as Ovaprim. This synthetic hormone is now replaced hypophysation technique, used worldwide as a best fish induce breeding agent. This product contains an analogue of salmon gonadotropin releasing hormone and dopamine antagonist. Doses used for various fishes for breeding. Only one time injection.

Males of Indian major carps and Chinese major carps– 0.2 ml/kg body weight and for females–0.5 ml/kg body weight.

Ovatide

M/S Hemmo pharma and scientist of Central Institute of Fisheries Education (CIFE), Mumbai, developed new synthetic hormone "Ovatide" for fish breeding in 1998, and successfully used for breeding of carps, catfishes, mahseer, etc. Doses for major carps males at the range 0.1 to 0.2 ml/kg body weight and for females in the range of 0.2 to 0.4 ml/kg body weight.

Breeding Devices

In corporation of various types of fish breeding devices in carp culture is in consequence of the development of induced breeding technique. The brief description and working of some common important carp breeding devices are given here.

(1) Hapa

A breeding hapa is a box-shaped rectangular container provided with ropes at its four top and bottom corners for tying with bamboo poles. All the sides of the hapa are stitched and closed excepting a portion at one end of the upper horizontal wall, which can be closed or opened, as and when required.

The injected breeders along with the hand-nets are introduced through this open end and the fishes are released into the hapa. The open flap of the horizontal wall of the hapa, after introducing the breeders is closed and tied properly to the lower vertical flap so that the breeders may not jump out and escape. For opening and closing the open end of the hapa, the lace-system or button-system or loop-system may be adopted to facilitate easy closing and opening of the flap. After spawning breeders are taken out and fertilized egg released into hatching hapa.

(2) Cement Cistern

Instead of the hapas, cemented cistern as big as the hapas can also be used for breeding. About 30-35 cm height of water is maintained in the cistern, preferably with circulation of water. Tap water or well water may be used, but chlorinated tap water must be avoided.

(3) Bangla Bundh

Bangla bundh is a dry type of cemented bundh-tank with specified slopes at sides and bottom. It is generally a masonry structure with arrangement of proper out letting of excess water and complete drainage. The side slopes of the tank are either brick-pitched with cement pointing or entirely plastered with cement mortar. Bottom of the bundh is cemented.

Bangla bundh can advantageously be constructed on any type of soil. The trapezoidal or rectangular shape of the bundh is found to be most ideal. The Bangla bundh made of any size depending upon the availability of land. However, a tank of 20-25 m length and 5-8 m width with an area of 0.01-0.02 ha is sufficient to accommodate 15-25 brood females of total weight of 40-50 kg and 20-30 mature males of total weight of 40-60 kg in every operation and can produce 4-8 million eggs at a time.

About 60 per cent of the area of the bundh, *i.e.* the narrower part is shallow and the rest is deep and broad. The zone nearer the outlet is the deepest for complete drainage of water by the gravitational force. Feeble water current is maintained.

(4) Chinese Circular Breeding Pool

The circular breeding pool developed in China during 1970. Therefore introduced in various modified forms in India. Principle of this hatcheries is to create artificially natural breeding conditions, like maintain continuous water current like natural riverine condition by gravity, raining condition by showers, cooling of water etc.

Showering over the spawning pool to increase the oxygen level of the water and to maintain the water temperature within a limited range. The spawning pool which appears like clock-shape structures, has two parts: a) a circular (6-8 m diameter and 1.2–1.5 m height) breeding chamber proper and b) a rectangular (3 x 2.5 x 1.6 m) egg collection chamber connecting each other by a 10 cm dia underground G.I. pipe. The water holding capacity of the circular pool is 35-70 m^3 at maximum level, but only 20-30 m^3 water is normally maintained for breeding, 25 to 40 sets of circular tank is 1:30 towards the orifice.

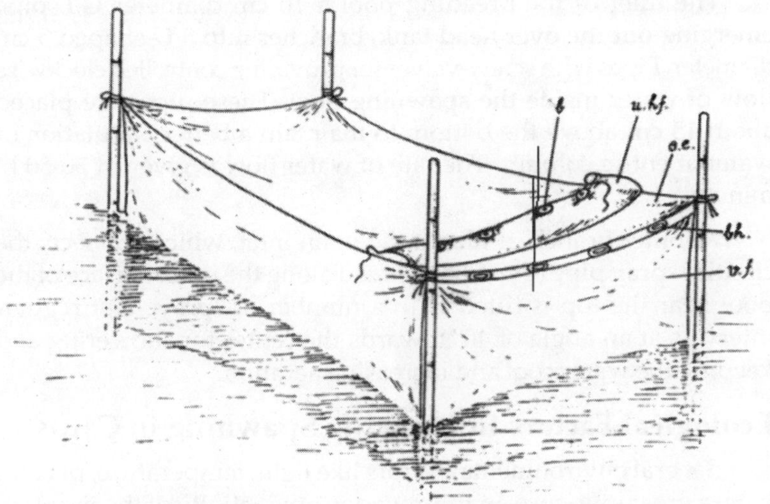

Figure 3.7: A Breeding Hapa, Showing a Portion of the Upper Horizontal Flap Open (The open end), Fixed in Water with Four Bamboo Poles; *b*: Button; *u.hf*: Upper horizontal flap; *o.e.*: Open end; bh: Button hole; *v.f.*: Vertical flap.

Figure 3.8: A Trapezoid-shaped Mini 'Bangla Bundh' in Operation

The inlet of the breeding pool a 10 cm diameter G.I. pipe emerging out the over head tank, branches into 3 L–shaped 5 cm diameter. Pipes with screw-valves for providing controlled clockwise flow of water inside the spawning pool. These pipes are placed about 15 cm above the bottom to maintain a better circulation of water at entire column. The rate of water flow is given at 30-50 l/ minute.

Another branch of line of the main inlet, which connects the circular spray pipe (4 cm dia.) spread along the inner surface of the pool near the top is fitted with a number of sprayers at regular intervals at an angle of 40° towards the center for showering and keeping the water cool and increasing aeration.

Ecological Factors Influencing Spawning in Carps

Several environmental factors like light, temperature, pH etc., have a great influence on the gonadotropic activity of the pituitary gland of fishes, and play an important role in stimulating the release of pituitary gonadotropins, thus controlling the reproduction in fishes.

Light

Light is an important factor in controlling the reproduction of fishes. Early maturation and spawning of fish as a result of enhanced photoperiods in *Notriops bifrenatus*, *Oryzias latipes* etc. Contrary to these there are observation on delayed maturation and spawning of some species of fish in the northern latitudes. In India *Cirrhinus reba* is observed to attain early maturity when day lengths longer than natural day. The resorption of gonads in this species is delayed and spawning condition can be maintained up to November. This prolongation of the spawning period is important from the viewpoint of pisciculture. These studies reveal that the requirements of light for activation of reproductive cycle vary from species to species and from place to place.

Temperature

Optimum temperature ranges for induced breeding of cultivated fishes and critical temperature limits, above and below which fish will not reproduce. The maturity is suppressed at low temperature in several species like *Fundulus*, Tilapia and *Perca fluviatilis*. On the other hand warm temperature stimulates maturation of gonads in

HORMONAL CONTROL OF REPRODUCTION IN FISHES

Environmental factors

↓

Hypothalamus

↓

Gonadotropin Releasing Hormone (GnRH)

↓

Pituitary gland

↓

Gonadotropic Hormones (LH and FSH)

↓

Gonads (Testis and Ovaries)

↓

Gonadal Hormones

↓

Maturation of gametes

↓

Spawning

Figire 3.9: Hormonal Control of Reproduction in Fishes (Environmental stimuli like photoperiod, temperature, etc. are received by exteroreceptor and reach the brain (hypothalamus). Environmental cues result in secretion of GnRH from the brain which in turns causes the release of pituitary GtH, GtH acts on the testis or ovary and stimulated the production of steroid hormone, 17α-20b-diOH Progesterone released from the testis and ovary induces the final maturation of germ cells leading to spermiation and ovulation.

several fishes and brings about spermiation. During induced breeding of Indian major carps spawning occurs at a temperature ranging from 24-34°C. Bhowmick (1969) observed spawning at 38°C in mahaseer (*Barbus putitora*). Temperature appears to serve as a triggering factor for spawning in carps.

Other Factors

Besides light and temperature several other factors like flood and rain, current of water, increased dissolved oxygen, shallow inundated spawning areas are important ecological factors for inducing natural spawning of carps in rivers and bundh type of tanks.

Spawning can be induced successfully in carps on cloudy rainy days, especially after heavy showers. According to Chaudhuri (1968) low pH, low total alkalinity, abrupt rise of water level and water temperature range of 27-29°C.

Chapter 4

Hatcheries and their Management

Developing eggs of selected species of carps are carefully transferred to small sized ponds called 'hatcheries', where they hatch. There are various type of hatchery. The commonly used hatching devices are given below:

(1) Hatching Pits

Small tanks, usually of 8′x 4′x 2′ are used for hatching fertilized eggs, and are located near the riverine collection grounds. A continuous but slow flowing water is desirable for aerating the eggs. A smaller cloth tank called hapa of 6′x 3′x 1.5′ of cheap coarse cloth and mosquito curtain cloth, is fixed up.

(2) Hapas

Fertilized eggs or developing spawn are transferred to small, triangular tanks made of coarse cloth, fixed in the river by means of bamboo poles. These tanks are called 'hapas' and each is about 90 x 45 x 30 cm, made of round meshed mosquito curtain cloth. This inner tank is fixed in a slightly larger outer tank made of coarse cloth. Several such hapas are fixed in flowing water of the river or pond by means of bamboo poles, and float freely so that eggs receive necessary aeration from water. A single layer of fertilized eggs are spread in the inner hapa to avoid over crowding. After hatching

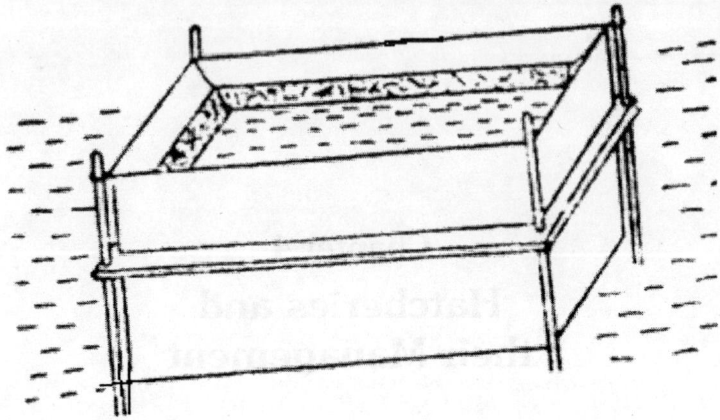

Figure 4.1: Open Hatching Hapa

hatchlings come out from the eggs. The hatchlings pass out through the mesh into the outer hapa, leaving the egg membrane behind, in

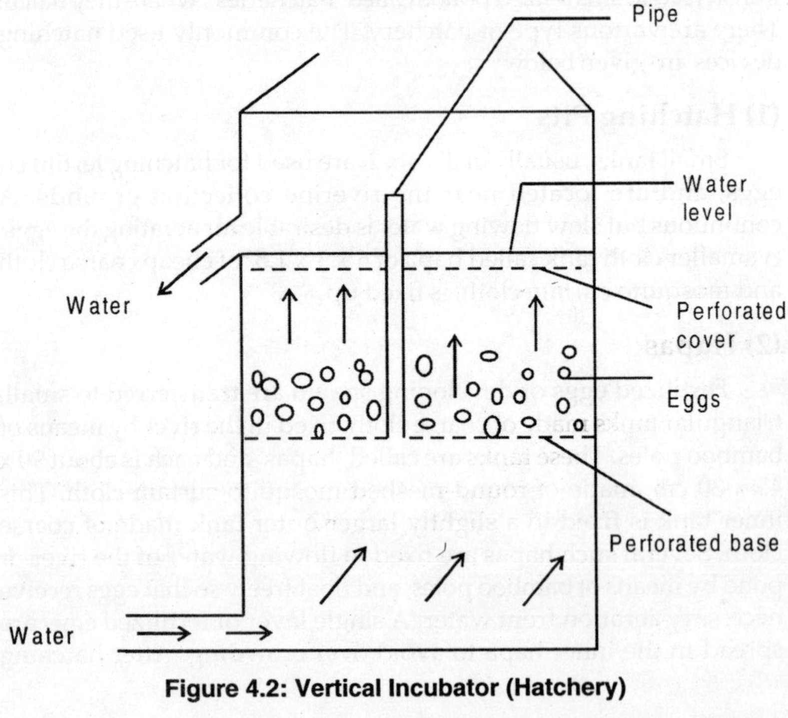

Figure 4.2: Vertical Incubator (Hatchery)

the inner hapa. The hatchlings are allowed to remain in the hapa for 36–48 hrs, after which they are transferred to the nurseries.

Hatching hapas are now being replaced with different types of vertical hatcheries made of plastic, Chinese hatchery pool, glass or polythene, and are used for hatching eggs in large numbers. These indoor hatcheries are more dependable and economical also.

(3) Chinese Circular Cistern Hatchery

The circular spawning pool developed in China during 1970, so it is also called Chinese Hatchery. Therefore introduced in various modified forms in India. Principle of this hatcheries is to create artificially natural breeding conditions, like maintain continuous water current like natural riverine condition, raining condition by showers, cooling of water, increasing oxygen level of water etc.

The Chinese circular hatcheries consist of different types of tanks and they are linked with each other: (*i*) Over head tank (*ii*) Spawning pool, (*iii*) Incubation or hatching pool, and (*iv*) Spawn collection tank.

(*i*) Overhead Tank

The overhead water storage tank circular masonry structure having a capacity of about 30,000-50,000 lit. Height 3.5 m above the base of the spawning pool. Filling of water in overhead tank by two separate G.I. pipes of 7.5 cm diameter. A temporary hut-shed or a permanent masonry structure may be provided over the tank to avoid direct sunlight and sudden rise of the water temperature. These water is used for fish breeding. Due to height of over head tank water come into breeding or hatching pool in forced.

(ii) Spawning Pool

The spawning pool which appears like clock-shape structures, has two parts: a) a circular (6-8 m diameter and 1.2–1.5 m height) breeding chamber proper and b) a rectangular (3 x 2.5 x 1.6 m) egg collection chamber connecting each other by a 10 cm dia underground G.I. pipe. The water holding capacity of the circular pool is 35-70 m^3 at maximum level, but only 20-30 m^3 water is normally maintained for breeding. The bottom slope of the circular tank is 1:30 towards the orifice. Each operation 25 to 40 sets of breeding pairs were released.

The inlet of the breeding pool is a 10 cm diameter G.I. pipe emerging out the over head tank, branches into 3 L–shaped 5 cm diameter. Pipes with screw-valves for providing controlled clockwise flow of water inside the spawning pool. These pipes are placed about 15 cm above the bottom to maintain a better circulation of water at entire column. The rate of water flow is given at 30-50 l/ minute.

Another branch of line of the main inlet, which connects the circular spray pipe (4 cm dia) spread along the inner surface of the pool near the top is fitted with a number of sprayers at regular intervals at an angle of 40° towards the center for showering and keeping the water cool and increasing aeration. About 70 kg weight of females can be released for breeding in pool and gives approximately 10 millions eggs. The fertilized eggs automatically goes towards the center of pool due to centrifugal force and through the out let pipe eggs collected in hatching pool.

(*iii*) Incubation or Hatching Pool

The incubation or hatching pool is a circular cemented tank measuring 3.0–4.5 m diameter and 1.2 to 1.5 m deep. The bottom slope from the periphery wall to the centrally placed orifice is 1:25. The orifice is a 10.0 cm diameter G.I. turn down outlet pipe. The incubation pool is conveniently constructed closer to the spawning pool and about 30 cm down below the latter to facilitate automatic transfer of eggs from the breeding pool to the hatching pool by slope gradient or at the same bottom level when the egg transfer is made by manual labour.

The incubation pool consist of two chambers the radius of the inner chamber is slightly larger than that of the outer one. Both chambers are partitioned by a monofilament cotton markin cloth of GARFAB (GHP 40-40/20) fitted on the vertically and horizontally arranged cross iron rods fixed on the 28 cm thick x 20 cm high circular wall of the inner chamber. The incubation pool is directly connected with the overhead tank by a 7.5 or 10.0 cm diameter G.I. pipe, which after being branched into two 5 cm pipes join the two underground circular 5 cm pipes. In the chamber 21-24 numbers of duck-mouth (each 2 cm dia) pipes fitted in two rows (such as 7-8 numbers on the inner and 14-16 numbers on the outer circular pipe) above the ground at 90° angle in regular intervals of clockwise

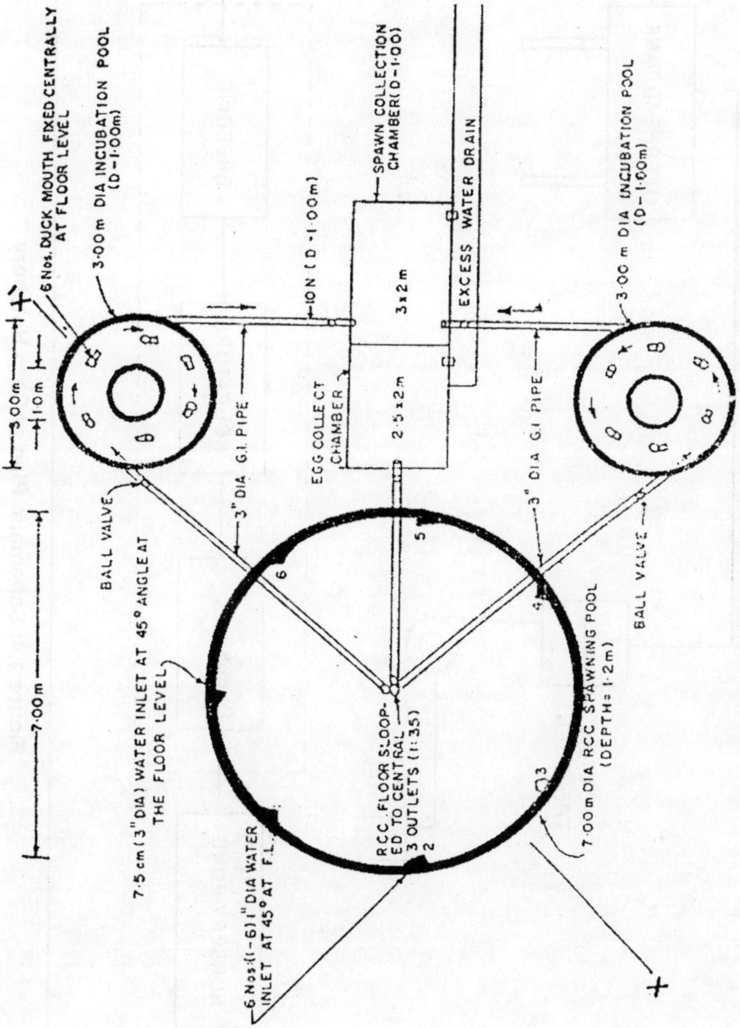

Figure 4.3: Plan of Carp Hatchery (Eco-hatchery)

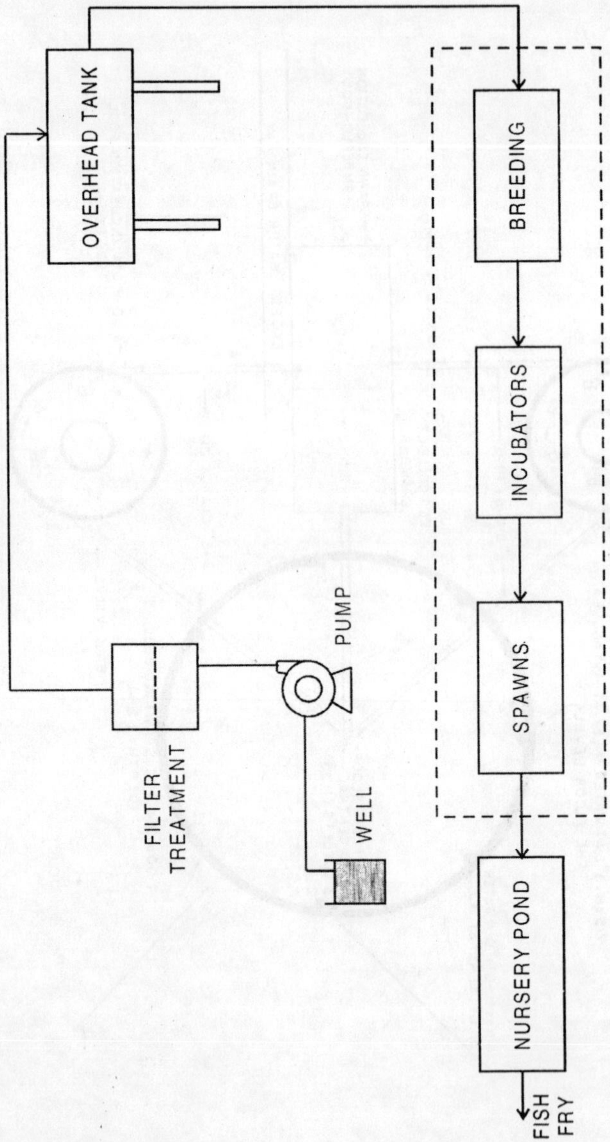

Figure 4.4: Schematic Plan of Carp Hatchery

direction one present to maintain a centrifugal force of water at desired speed (45-60 l/min) for sieving out the disintegrated egg shells after hatching through the partition cloth into the inner chamber and there after through the outlet pipe.

(*iv*) Spawn Collection Tank

The spawnery a rectangular cemented tank (dimension of 3 m x 2 m x 1 m or more or less) is constructed adjacent/attached to the incubation pool. A spawn collection pipe of 10.0 cm diameter, fitted with screw valve.

Management of Hatchery

After each breeding operation spawning pool is washed with dilute potassium permanganate solution to disinfect and then washed with water and then kept dry for some time. For washing use the rough brush. After complete washing and drying. Then pool is used for next breeding operation. Similar way like spawning pool the washing and drying of the incubation tank after each operation is done. In each operation alternate use of incubation pool. Cloth hapas should be washed and dried after every 2 or 3 days' use and, therefore, there should be spare cloth hapas also.

Chapter 5
Rearing of Spawn and Transport of Fish Seed

Fish seed is the term used to include various developing stages of a fish, required for stocking ponds for culture. This includes spawn, hatchlings, sac-fry, fry, juveniles, fingerlings and yearlings. Majority of fish seed collected at the spawn stage. Spawn is a fertilized developing ova. The sources of obtaining spawn by two ways: one is from the natural source *i.e.* rivers and streams and from bundh spawning and the other is the spawn from artificial breeding of fish *i.e.* stripping method and the pituitary hormone induced breeding. Spawn collected from the rivers and wet bundhs, besides comprising a mixture of numerous species, usually has large sized associates, many of which are predaceous. The spawn is, therefore, sieved through a round meshed mosquito netting cloth on to a muslin cloth suspended in water in order to separate the different species.

After spawn collection there rearing is important. The spawn are reared in nursery pond, which is size approximately 50′x 50′x 4′. Before releasing the spawn into nursery, the preparation of nursery is an important step for successful rearing of carp spawn to fry. The occurrence of large scale mortality of spawn stocked in unprepared nurseries is a common experience of fish culturists.

Preparation of Nurseries

Efficient pond fish culture entails special preparation of nurseries to receive the tender hatchlings and spawn. Small and seasonal nurseries are preferable. The drying up water bodies in summer months helps mineralization, removal of excess of organic matter and automatic destruction of predators, minnows, etc., that usually abound in perennial nurseries. The steps commonly adopted towards preparation of nurseries are:

1. Control of predatory and weed fishes
2. Fertilization of ponds aimed at sustained production of zooplankton to serve as food for spawn
3. Control of algal bloom, and
4. Eradication of predatory insects.

Predatory fishes breed in ponds their young ones feed vigorously feed on the plankton which is main food of spawn. Weed fishes and their young ones directly feed on carp hatchlings and spawn. Removing unwanted fishes from nursery ponds by repeated drag netting for certain bottom dwelling fishes repeated seining for deeper waters used hooks and lines to catch large predators. Dewatering of nursery is also another way but this is not possible every time. The nursery poisoned for removal of unwanted fishes is the best way. Derris powder, with 5 per cent rotenone content is perhaps the most commonly used fish poison before one month release of spawn in nursery. A dose of 1 ppm is recommended to kill predatory fish and insects.

Manuring of nursery ponds for the production of zooplankton is essential. The natural food of spawn for this, the nursery ponds are first limed just after removal of unwanted predatory and weed fishes. After liming nursery ponds are treated either with organic manures such as cow dung or inorganic fertilizers (N.P.K.) cow dung 10,000 kg/ha, before 15 days of spawn release, second application 5,000 kg/ha, 7 days after stocking, is to maintain sustained production of zooplanktons.

For removal of aquatic insects or killing, spraying oil to kill the insects. Carp spawn, immediately after stocking need a plentiful supply of natural food in the form of animalcules. For maximum survival of spawn, only such nurseries should be selected as contain rich crop of zooplanktons. A minimum 1.5 to 2.0 ml zooplankton in 54.5 litre of water is essential.

Stocking of Spawn

Depending upon the density of plankton, the stocking rate of spawn varies from 12–20 lakhs/ha for 14-15 days. In order to know the quantity of spawn for stocking the spawn should be carefully measured in graduated metal cups with perforated bottoms and put in a container to avoid any abrupt change in water quality and temperature, the spawn to be stocked should be kept in a suitable container in the water of the transporting medium to which the pond water should be gradually added.

The container should then be slowly dipped and tilted in the pond so that spawn are free to swim out of container in the nursery. It is better to release spawn in different section of nursery. The stocking of nursery pond is best done late in the evening, a procedure which gives the spawn time to acclimatize themselves during the ensuing night.

Supplementary Feeding

Soon after being stocked, spawn start feeding voraciously on zooplankton. At this time, the feeding requirement of spawn are so large that within 2-3 days of stocking, the food available in the nursery become very low. For the first two days after commences the spawn do not take artificial feed but subsist almost completely on natural food. After this period artificial feed along with the natural food in enhances the growth and survival of spawn considerably. Powdered oil cakes and rice bran, which are given as artificial food to spawn should be shifted through a fine-meshed sieve to remove the husk from rice bran and to ensure uniformity of particle sizes. Daily dose of artificial feed should be related to the weight of the spawn stocked. Hatchlings of 6 to 6.5 mm length on an average weight c 1.4 mg. Spawn is fed four times its initial weights at the time of stocking for the first 5 days followed by 8 times the initial weight for the second 5 days period. No feeding is done on the 11th day for conditioning and the fry are harvested on the 12th day. Spawn stocked in nursery ponds ordinarily attain a length of 22-25 mm in 15 days with artificial feeding giving survival of more than 50 per cent. Then the fry are netted out and released in rearing ponds for further development.

Transport of Fish Seed

Fish seed collected in large number either from rivers or from the sources and are transported to the market for sale and then to the nurseries of fish farm. Transportation of fish seed is considered to be a crucial step in aquaculture, because mishandling of seed causes heavy mortalities. To avoid this mortality, the transport of seed should be properly planned and implemented. "Fish seed" is the term used to include various developing stages of a fish, required for stocking ponds for culture. This includes the following:

1. Spawn–developing fertilized ova.
2. Hatchlings–stage immediately after hatching.
3. Sac fry–newly hatched larvae with yolk sac.
4. Fry–young larvae after yolk sac is absorbed.
5. Juveniles–young ones resembling adult fish.
6. Fingerlings–young ones of large fishes, resembling the adult fish, but only 4-12 cms in length.
7. Yearlings–young ones that are one year old.

The fish seed mainly consist of spawn, fry, and fingerlings. Fish seed can be obtained from two sources: (i) by natural breeding of fishes in rivers and streams, and (ii) by artificial breeding of fish. Fish seed are collected in large number either from rivers or from other sources and are transported to the market for sale and then to the nurseries of fish farms. Transportation involves heavy mortality, as a very large number of seed are held in a small quantity of water for a certain period of time. The main causes of mortality of all stages of fish seed during and immediately after transport are:

1. Deficiency of oxygen in water or high carbon dioxide content.
2. Increase in temperature of water of the container.
3. Toxicity due to accumulation of waste like ammonia.
4. Injury and infection during transport.
5. Hyperactivity, strain and exhaustion of the fry.

The problems of transport depend on the fish species, age and size of the fish to be transported, distance to be covered, speed of transport, and the type of container used for transport. At recent times several methods are now available for easy and safe transportation of fish seed. However, these problems can be reduced to a considerable degree if certain precautions are taken, as mentioned below:

Conditioning

A common method of conditioning is to store the fry in a cloth hapa in ponds or in a still part of the river. Fry are subjected to 'conditioning', to acclimatize them to a limited volume of water and to empty the gut. For this, a number of fry are kept in a limited volume of water, usually in a cloth hapa. As conditioning involves starvation, it should not be prolonged, and the period depends on the size and health of fry. The carp fry are generally subjected to 48-72 hrs of conditioning. The primary purpose of confinement and starving is to force the fish to remove all faecal matter before being transported, since this faecal matter quickly contaminates the water and reduce the content of available oxygen. Clean water at about 20.5°C is considered suitable for conditioning. During conditioning and transportation, fry and fingerlings should not be handled with bare hands as a removal of slime and scales of the body may result in fungal and bacterial infections. After proper conditioning, fry are transported either in open or closed vessels.

Transportation

There are two methods of transporting the fish seed.

1. Open system comprising open carriers, with or without oxygenation/aeration.
2. Closed system having sealed air-tight carriers with oxygen.

Transportation of fry and fingerlings in open earthen vessels called 'hundies' has been the traditional method practiced in West Bengal. Hundies are of various sizes, smaller ones for transportation on heads, and larger one by rail. These vessels are filled with water from the same source as the fry. A vessel of 20-25 litre capacity is stocked with 50,000 fry, while 75,000 are kept in a vessel of 35 litre. A small quantity of alkaline soil is added to the vessel to buffer the pH fluctuations. The vessel is periodically shaken for aeration, and if necessary water is partially renewed at intervals during transit. Fry coming up and grasping is a sign of oxygen deficiency, and indicates need for changing the water. Dead fish have to be removed immediately to avoid pollution. The temp of water can be kept down by a wet cloth covering over the carrier. But these traditional methods often heavy mortality during transport.

Now large sized aluminum vessels with wide mouth are used in place of hundies. The mouth is closed by a perforated lid. In Orissa, fry and fingerlings are transported in tins. These vessels are transported in open motor vans, fitted with hand pumps to circulate water. The modern transportation devices include closed systems like plastic bags, bins, buckets, collapsible plastic pools and fiber tanks. But, most convenient way of transportation of fish seeds is the use of oxygen-packed polythene bags (Figure 5.1). Spawn and fry are now transported in sealed metal containers with oxygen. Specially designed containers of aluminum, or galvanized iron or polythene bags are used for the purpose. Polythene bag of 76 cm x 45 cm and of 33 liter capacity held in a tin container (18 litre capacity)

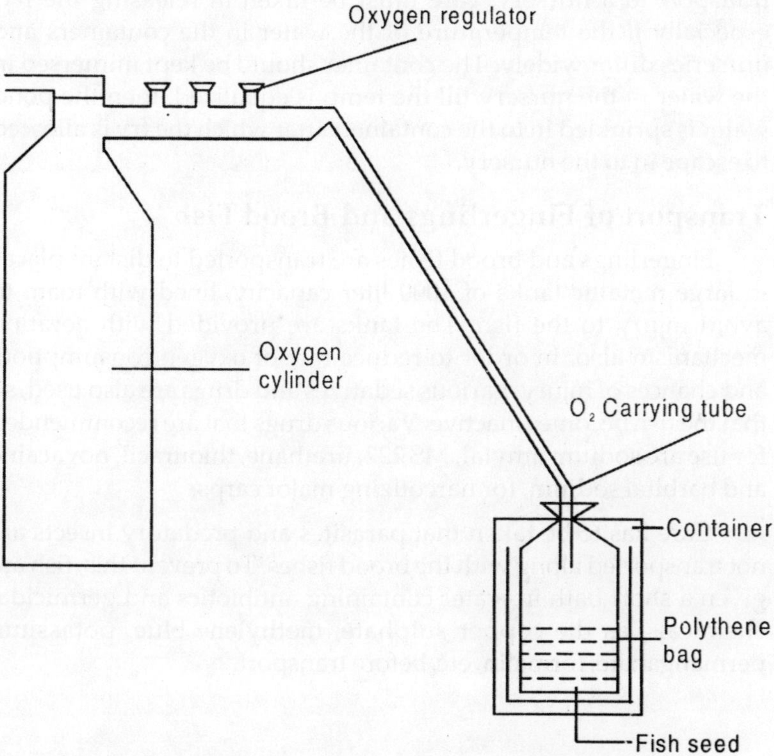

Figure 5.1: Apparatus for Oxygen Packing Seed

are now widely used in India for transport of major carp seed under oxygen. A layer of corrugated paper packing is inserted in between the two, for insulating against external heat. About 6 to 7 liters of cool water devoid of plankton and insects, is filled in the bag. Cool water reduces the activity of fish, decreases the need of oxygen and holds more dissolved oxygen. Transportation of fish seed always done with the coolest water available. Conditioning fish seed are put in the bag in a quantity determined by the size of fish seed. About 1100 fry measuring 16 to 20 mm in length in one bag for 24 hrs of transportation.

Other Precautions

Transport of seed is under taken during night or in the early morning to avoid temperature fluctuations. Warming of water is prevented by covering the containers with wet cloth or ice. After transport to a nursery, care must be taken in releasing the fry, especially if the temperature of the water in the containers and nurseries differ widely. The container should be kept immersed in the water in the nursery till the temp is equalized, then the pond water is sprinkled in to the container, after which the fry is allowed to escape in to the nursery.

Transport of Fingerlings and Brood Fish

Fingerlings and brood fishes are transported to distant places in large metallic tanks of 1000 liter capacity, lined with foam to avoid injury to the fish. The tanks are provided with aeration mechanism also. In order to reduce rate of oxygen consumption, and chances of injury, various sedatives and drugs are also used, so that the fish becomes inactive. Various drugs that are recommended for use are sodium amytal, MS 222, urethane, thiouracil, novacaine and barbital sodium, for narcotizing major carps.

Care has to be taken that parasites and predatory insects are not transported along with the brood fishes. To prevent this, fish are given a short bath in water containing antibiotics and germicidal chemicals, as the copper sulphate, methylene blue, potassium permanganate, formalin, etc. before transport.

Chapter 6

Design and Construction of Freshwater Fish Farm

In piscicultural programmes, fish culturist requires different types of ponds rearing various sizes of fishes in fish farm. Different types of ponds to be digged for rearing fishes. For carp culture, at least three different types of ponds, *i.e.*, nursery, rearing, and stocking ponds are required. All these different kinds of ponds differ from each other with respect to its size, area and depth of water. The nursery ponds are small and shallow having an area of 0.02 to 0.08 ha (20 m x 10 m to 40 m x 20 m in size) with a depth of 1.2 to 1.5 m. These ponds are used for raising fry from spawn (5.6 to 25-30 mm). The rearing ponds are slightly larger in size than the nursery ponds. The pond size vary from 0.08 to 0.10 ha (40 x 20 m to 50 x 20 m) with a depth of 1.5 to 2.0 m. These ponds are used for rearing fingerlings from fry. The stocking ponds are larger than rearing ponds. They differ in size from 0.2 to 2.0 ha with a depth of 2.0 to 2.5 m. These ponds are utilized for stocking fingerlings in order to obtain table size fish. All the ponds in general should be rectangular in shape. The pond width should not exceed 40 m. This will facilitate longer run for fish to achieve healthy growth and also smaller nets and lesser number of people are required to catch the fish.

The number of likely factors to be taken into consideration for farm pond construction can be broadly divided into 3 categories.

1. Site selection
2. Designing and
3. Construction

Site Selection

The primary consideration in constructing a fish farm in the selection of a site which would provide good quality of water retensive soil base and pond sites, having an assured adequate water supply to fill in ponds at any time of the year. The economics of fish pond construction and the productivity of pond largely depend on the selection of suitable site for culture activity. The more important aspects of fish farm construction are topography, soil type and water supply.

Topography

Topography or layout of land is the surface feature of water shed. The ideal topography of a farm site is gently sloping terrain of a wide valley or basin surrounded on three sides with high lands and with a narrow outlet on the fourth.

Soil Type

The soil should be free from vertical and lateral seepage. Such impervious soils are ordinarily heavy clay, silty clay, clay loam, etc. Ponds should not be constructed over sandy soils because sand is too porous to retain water. If the selected site is poor in water retentivity, clay has to be brought from elsewhere and a clay coating and puddle core has to be provided to make the ponds retain water. Large quantities of water and cow dung help in improving the water retention capacity of poor soils.

Water Supply

The third requisite for selecting a site is the availability of an adequate and dependable source of water. The usual sources of water supply are: reservoirs, streams, springs, canals, surface runoff and wells. A source of water supply to fill the ponds against losses due to seepage and evaporation is there fore essential and should be ready at hand.

Designing

Construction of Fish Ponds

Any suitable area can be converted into an earthen pond by digging 0.5 to 1 m deep. Rectangular pond are suitable than square or circular ponds for easy netting operations. A typical earthen pond will have the following structures, *viz.*, bunds or dikes, harvesting pit, inlet, outlet and core trench.

Table 6.1: Requirements of a Normal Fish Farm Site Under Indian Conditions

Particulars	*Requirements of Normal Fish Farm Site*
Nature of terrain	Non rocky with at least 2 m deep soil
Slope of the terrain	Land should be level or gently slopping
Physical quality of the soil	Soil fraction should be about 90 per cent of the whole soil, stone and gravel not exceeding 10 per cent
Chemical quality of the soil	pH near neutral. Total nitrogen > 0.1 per cent, phosphorus > 0.1 per cent, organic carbon > 1.0 per cent, free $CaCO_3$ > 5 per cent
Water supply	Perennial water supply nearby
Biological productivity	Average plankton production per m^3 should range between 10 and 20 ml

Bunds

Bunds are the protecting structures of fish ponds and are essential for both drainable and undrainable fish ponds. They may be of three types, *viz.*, a) main bund or peripheral bund which is essential for large fish farms enclosing a number of fish ponds of varied sizes, b) bunds holding water on one side and, c) bunds dividing two adjacent ponds.

The main bund is the highest part and is constructed in the deepest part of the pond area. The bund holding water only on one side is also constructed as main bund, but it is not constructed on the deepest area of the pond.

Slope

The life span and strength of any bund depend not only on the quality of soil but also on its slope and crown width or crest. The slope may be defined as the ratio of horizontal increase in the base of the bund from the point of perpendicular to the top edge of the bund to the edge of the base of the bund on the same side per unit length. The slope of a bund is based on the quality of soil used for its construction and required height. Normally, good clayey soil makes steeper slope than the sandy-clayey or sandy soil. For ordinary earthen ponds of less than 0.5 ha, the wet side slope may be 1 : 1 : 5 and the dry side slope 1 : 1. The crown width also depends on the height of the bund. However, a minimum of 1 m crown width is invariably needed for any bund of a fish farm is to be motorable one, a minimum crown width of 3 m is essential.

Berm

If the area of the pond is more than 0.5 ha, a platform-like space between bund and watery area known as berm or bench line should be made available. The width of the berm may vary between 0.5 to 1.0 m, depending on the height of the bund and the size of the pond. The berm apart from serving as a walkable space for fisherman also protects the bund from direct contact with water.

Construction of Pond

It will always be economical if the bund is formed out of the earth removed from the enclosing area. Hence, before constructing any bund, it is better to calculate the total quantity of the earth required for constructing bunds for all the four sides of a fish pond and the area to be dug in the enclosed area for getting the earth.

Determination of Quantity of Earth for the Construction of Bund

To find out the quantity of earth required for the construction of bunds, it is necessary to determine the quantity of earth required for one meter long bund with the proposed specifications regarding slopes, height, crown width, etc.

Total length of actual four bunds (20+50+20+50) = 140 m. Total earth required for 140 m long bund = 140 x 2.25 M^3 = 315 m^3. In order to form the four bunds of the above specifications, it is enough, if earth is taken from a depth of 0.315 m from the enclosed area, *i.e.*, 1,000 m^2.

Bund Formation

Sandy or gravelly soil is not suitable for bund construction. Further, a soil with high clay content may also form fractures in the bunds. Hence, a mixture of silt, sand and clay in the ratio of 1:3:2 should be used in the construction of firm bunds. After all the area data relating to the bund are gathered, a design is made for the pond construction. Using the design, the bund lines are pegged along

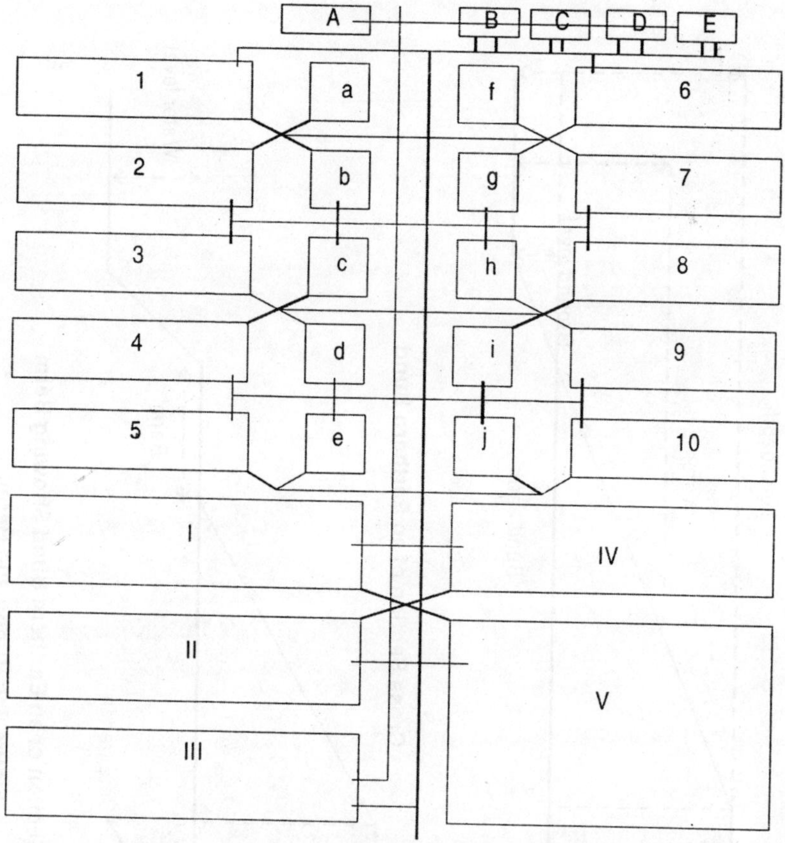

Figure 6.1: Carp Fish Farm
A-E: Hatching pits; a-j: Nursery ponds; 1-10: Rearing ponds; I-V: Stocking ponds

with the contour of the pit from which the required earth is taken out. The axis of the bund with a base of known length is first pegged and their sides are demarcated. After pegging the bund line, the debris, roots, stones etc., are removed from the marked area. Subsequently, the top fertile soil of 5-10 cm depth is also removed. This may be heaped in a nearby area for future use. If the soil of bund area is of pervious material, then the formation of a core trench is a must. The core trench depends on the base width and height of the

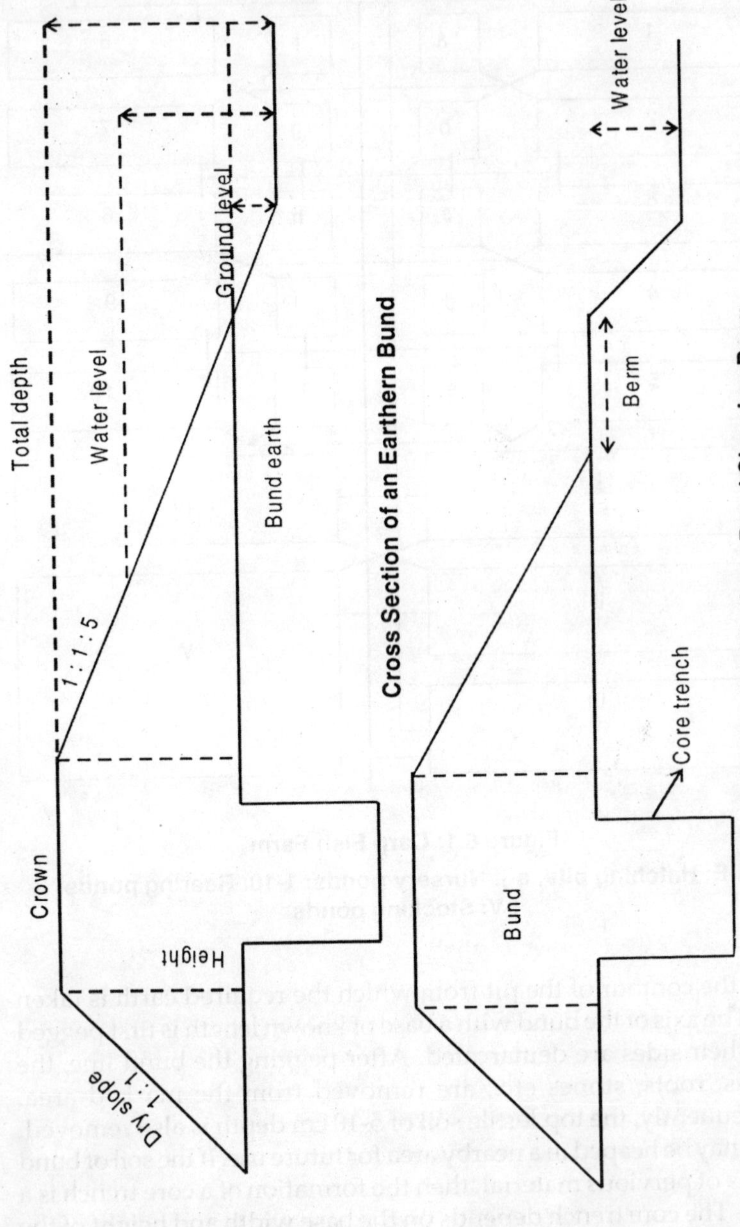

Cross Section of an Earthern Bund Showing Berm

Figure 6.2: Bundh and Berm

bund. For example, a bund of 3.5 m base width may have a core trench of 0.75 m width and 0.75 m depth. After the area pertaining to core trench is marked, it is dug and the brought from outside is packed in layers and compacted. The core trench is mainly to prevent seepage of water through bund and to strength it, and it may be avoided if the pond is not deeper than 1-1.5 m and soil is clayey and impervious. The profiles of the bunds are then made with long and thin wooden pieces, which may reflect the height and crest of bund. The final phase of the bund construction is fitting out, where the bits of earth are properly packed according to the bund profile setup. Subsequently, the top soil heaped apart is spread over the slope which would not only enhance the grass production but would also avoid the erosion of the bund. To hasten the growth of grass, adequate amount of turf may also be placed on the slope. Finally, the pit should be leveled. Before filling the pond with water, its bottom should also be leveled and cleared of debris, stones, etc. A gradual but insignificant slope from one end to the other end in the case of undrainable fish pond may be provided at the bottom. This will help, particularly during summer or harvesting period, when water level goes down considerably and the harvestable fish can find refuge in the deepest area with water.

Inlet and Outlet

These are required especially for larger ponds, in order to ensure a smooth water supply and drainage. Further, they are helpful in preventing the entry of wild fish from outside and escape of the fish from inside the pond. For pond of 0.1- 0.2 ha area, an inlet or outlet pipe of about 15 cm diameter will be ideal, as a pipe of less than 10 cm may lead to clogging. The outlets or drains are invariably installed at the lowest plane of the pond. However, for larger ponds (1-2 ha) the pipe should be of around 25 cm diameter. The inlet/outlet pipes are to be provided with suitable screens, so as to keep off wild fish as well as prevent escape of cultivated fish. However, to stop the flow of water through the inlet and outlet pipes, in small and medium sized fish ponds, plugs of wooden slumps or gunny cloth also be used as and when needed. For larger fish ponds the construction of concrete inlets and outlet are highly useful.

While designing the fish farm with nursery, rearing and stocking ponds the following points should be borne in mind in addition to the survey reports and maps.

In case of fish farm is meant for fish seed production only, nursery and rearing ponds can be constructed leaving a normal area for stocking the breeders. If the farm is aimed to produce the table size fish then, only the stocking ponds needs to be constructed. All pond should be rectangular in shape and the width of the pond not exceed 40 m. Embankment of the fish pond should be designed in such way that its height from the bottom of the pond should be 2.7 to 3.7 m. A 0.6 m free board should be left between the water surface and crest of the embankment. The free board is essential to prevent the pond spilling over due to sudden and exceptional downpours. A fencing is put around the fish farm in order to keep live-stock off the embankments and shore line areas of ponds. The cattle, if they trample sod, weaken the embankments. They make the water muddy if permitted to wade into the ponds.

Chapter 7

Physico-chemical Properties of Freshwater

Lakes, reservoirs and ponds constitute a great source of inland fisheries in India. Lakes are naturally formed depressions on the surface of the earth and get filled with water, while reservoir is artificially created body of water. The productivity of a pond or a lake depends up on the quality of water. The primary requisite for fish culture is the water, which for successful rearing of fishes should offer the most favorable conditions for to existence not only of fish but of other aquatic organism as well, which form the food of the fish. The principal physical and chemical condition of the natural water necessary for fish culture are given below:

Physical Properties of Freshwater

Depth, temperature, turbidity and light constitute the more important physical parameters on which the productivity of a pond depends.

Depth

Depth of a pond has an important bearing on the physical and chemical qualities of water. A shallow pond allows the sunlight to penetrate up to the bottom and increases productivity by photosynthesis. But if the pond is to shallow, water gets heated up during summer months affecting the survival of the fish. Generally 2 meters depth of ponds are considered good for maximum biological productivity.

Temperature

All organisms, including fish, possess well defined limits of temperature tolerance. All metabolic and physiological activities and life processes such as feeding, reproduction, movement and distribution of aquatic organisms are greatly influenced by water temperature. The oxygen content of water, in general, decreases with rise in temperature. Fishes of temperate regions can survive under ice in winter but those of the tropics cannot stand such low temperature. *Tilapia mossambica*, which belongs to the tropics, cannot stand temperature below 8.9°C. The Chinese carps belongs to the cold rivers of China, prefer temperature below 30°C. Indian major carps thrive well in the temperature range 18.3–37.8°C. Temperature below 16.7°C and above 39.5°C prove fatal to them. Upper lethal temperature of air breathing fishes, lies between 39–41°C. Hence a prior knowledge of the temperature range is necessary before introducing fishes for culture in a pond.

Turbidity

The turbidity of water may be either due to suspended inorganic substances such as silt and clay or due to planktonic organisms. It is an important limiting factor in the productivity of a pond. Water becomes turbid during rains and floods also, restricting penetration of sunlight and reducing photosynthesis. Respiration is effected and the gills of the fish are injured and get chocked due to large amount of suspended matter. During heavy floods, turbidity of the rivers like the Ganga, rises to 2000–5000 ppm. However, Indian major carps are able to tolerate to able up to 2000 ppm or even more. The suspended particles causing turbidity may also adsorb considerable amount of nutrient elements like phosphates, potassium and nitrogen making them unavailable for plankton production.

Fishes permanently in turbid waters show various adaptations like reduced eyes and the mucus secreted by their skin is capable of coagulating suspended matter to protect the gills from being chocked by suspended particles.

Light

Light is an important factor influencing the productivity of the pond water. Light penetration depends upon the available intensity of the incident light which varies with the geographical location of the pond. Shade provided by the surrounding trees on the shore and turbidity of water affect the incidence of light on the pond.

In shallow ponds, light is able to penetrate up to the bottom, and causes heavy growth of vegetation, occupy the entire basin. Advantage of shading effect if often taken in pisciculture for the control of algal blooms and submerged weeds. Light control the photosynthesis activity of the plants, thus indirectly affect in the oxygen content of the water of the pond.

Chemical Properties of Freshwater

The productivity of a pond depends on the presence of several chemical substances of which dissolved gases like oxygen and carbon dioxide, inorganic salts, pH and nutrients constitute the more important chemical characteristics of a pond.

Dissolved Oxygen

This is one of the most important factors limiting productivity in pond water. Dissolved oxygen is most important for the animal and plant life in a pond. It is available to the water by absorption from the surface and by photosynthesis of plants.

Cultivable fishes vary with respect to their oxygen requirement for aquatic breathing. For example, at 20°C salmonids need 9 mg/ litre strength of dissolved oxygen in water, but carps only 6-7 mg/ litre. Silurids and cichlids need lower concentrations. When water is having lowered concentrations of oxygen, fish begin to rise to surface.or crowd near inlets. In extreme depletions of dissolved oxygen level, fish may die of asphyxia.

Variations in concentration of dissolved oxygen may occur due to several parameters, i) cloudy days reduce it. ii) higher temperature decrease it, iii) higher salinity lowers it iv) decomposing organic matter, sewage fertilizers and manures may also bring about de-oxygenation of water.

During day time, generally, the oxygen concentration goes up, it goes down considerable during night. This diurnal version may in some cases lead to dangerous for in the level during night in ponds with excessive abundance of algae and submerged weeds, or even during day in cloudy day.

Carbon Dioxide

Large amount of carbon dioxide is available to the water through respiration. It is also derived from atmosphere as well as by decomposition of organic matter. Presence of carbon dioxide is essential for photosynthesis, but excess is harmful to animals. Carbon dioxide reacts with water forming carbonic acid (H_2CO_3) which dissociates into H^+ and HCO_3^- ions. In the presence of carbonic acid, lime-stone ($CaCO_3$) dissolves forming calcium bicarbonate Ca $(HCO_3)_2$, which influences the organic communities of water.

The importance of rain in supplying carbon dioxide to inland waters lies in the fact that CO_2 gets dissolved in rain water. In the deeper strata of mineral soil, it generally encounters limestone ($CaCO_3$), a calcium-bearing rock, which readily dissolves as calcium bicarbonate in the presence of H_2CO_3. The solution of $Ca(HCO_3)_2$ remains stable only in the presence of certain surplus amount of free CO_2. The amount of gas increases very rapidly with increasing lime content. The free carbon dioxide which is necessary to retain calcium in solution in the form of calcium bicarbonate is called the equilibrium or free carbon dioxide. The carbon dioxide is contained in two states-(i) as half bound (HCO_3) and (ii) as bound (CO_3), both together being called combined carbon dioxide. The dissolution of $CaCO_2$ is dependent upon the availability of CO_2 to the extent it is greater than that of free CO_2, this additional CO_2 being known as aggressive carbon dioxide. If a solution of Ca $(HCO_3)_2$ loses the equilibrium CO_2, the following reaction takes place: Ca $(HCO_3)_2$ = $CaCO_3 + H_2O + CO_2$.

This dissociation proceeds with the precipitation of $CaCO_3$ until there is sufficient CO_2 in the solution to bring about equilibrium. Various reactions involving CO_2 in natural waters are relate acid base relationships in the medium. Under acid conditions, the combination of CO_2 with H_2O is as under: $CO_2 + H_2O \leftrightharpoons H_2CO_3 \leftrightharpoons H^+ + HCO_3 \leftrightharpoons H^+ + CO_3^-$ and under highly basic conditions the reaction is:

$$\text{Base OH} + H_2CO_3 = H_2O + \text{base}^+ + HCO_3^-$$

Calcium bicarbonate, in a very dilute solution, is almost completely dissociated:

Ca $(HCO_3)_2 \leftrightharpoons Ca^{++} + 2HCO_3$–. Next step, in dissociation is $HCO_3^- \leftrightharpoons H^+ + CO_3^-$, which takes place in higher *p*H values. Carbonic acid is hydrolyzed as under:

$$HCO_3^- + H_2O = H_2CO_3 + OH^-$$

A solution of calcium bicarbonate has: (a) free CO_2 (equilibrium CO_2), (b) carbonic acid partly dissociated (HCO_3^-, H^+) and partly undissociated calcium bicarbonate (Ca^{++}, HCO_3^-, CO_3^-) and (d) hydroxyl ions resulting from the hydrolysis (OH^-).

$$Ca\,(HCO_3)_2 + KOH = CaCO_3 + KHCO_3 + H_2O$$

The above-stated reactions, caused by the presence of carbonate-carbonic acid mixtures, prevent major fluctuations in the alteration of *p*H of pond waters and create relatively more stable conditions for sustenance of animal and plant life. These mixtures of weaker acids and their salts are called *buffer*.

Total Alkalinity

Total alkalinity of the water is mainly caused by the cations of Ca, Mg, K, NH_4 and Fe combined either as carbonates and/or bicarbonates or occasionally as hydroxides. Total alkalinity has been variously termed as alkalinity, titrable base, methyl orange alkalinity, alkaline reserve or acid-binding capacity. Swingle (1967) suggested a rejection of all other terms in favor of total alkalinity of universal adoption. He suggested the titration end points to correspond to the following *p*H values:

*p*H = 5.1 when total alkalinity is 30 ppm as $CaCO_3$

= 4.8 when total alkalinity is 150 ppm as $CaCO_3$

= 4.5 when total alkalinity is 500 ppm as $CaCO_3$

Hydroxide alkalinity generally occurs in polluted waters. In other waters it is occasionally encountered during mid-afternoon in surface layers in waters showing intense photosynthesis. Total alkalinity, originally used by Schaperclaus (1933) as a measures of productivity, is believed to be so by many workers. According to Schaperclaus, the most productive water is that which titrates 2 ml to 5 ml of N HCl per litre or 200-500 ppm equivalent $CaCO_3$. Alikunhi (1975) stated that in highly productive waters, the alkalinity ought to be over 100 ppm. Many rivers and fish ponds fall within the total alkalinity values equivalent to 10-50 ppm $CaCO_3$. Waters of hilly streams; sandy, rocky or very clayey areas; flooded rivers in rainy season; waters in heavy rainfall areas and waters infested with submerged weeds usually have low total alkalinity values.

*p*H

The hydrogen-ion concentration of natural waters is an important environmental factor, the variations of which, among other causes, are linked with the species composition and life processes of animal and plant communities inhabiting them. The *p*H of the water is defined as the logarithm of the reciprocal of hydrogen-ion concentration. It may be expressed as mathematically as follows:

$$pH = \log \frac{1}{(H^+)}$$

(where, (H^+) is the amount of hydrogen ions in a solution in moles per litre). There is 0.0000001 of hydrogen ions and a corresponding quantity of (OH^-) ions in a litre of pure water. The *p*H of neutral water is, therefore, 7. Increase in the concentration of H^+ ions results in a lower *p*H value, or conversely, reduction in the H^+ concentration brings about a higher *p*H value, solutions are acidic in *p*H values below 7 and alkaline in values above 7. The pond water shows diurnal fluctuation of *p*H, being most alkaline in mid-afternoon, and most acidic in the early morning. Pond water may be alkaline, acidic or neutral, and is an important environmental factor influencing the species and metabolism of all animal and plants inhabiting it. Low *p*H is harmful to fishes. According to Swingle (1967), pond water having *p*H values from 6.5 to 9.0, is most suitable for culture. Water having values of *p*H below 5 and above 9.5 are not suitable. Fish dies at about *p*H 11. Acid water has following adverse effects: (*i*) it reduces growth rate by decreasing the appetite of fish, (*ii*) it increases chances of microbial infection, occurrence of diseases and attack of parasites, (*iii*) it has toxic effect, (*iv*) it lowers dissolved carbon-dioxide content, reducing photosynthetic activity, (*v*) it removes lime from water, thereby affecting growth of skeletal tissues and scales, (*vi*) The toxicity of H_2S, copper and other heavy metals to fish is increased by lowering the *p*H.

It has been found that slight alkalinity brings about *p*H stability, and therefore supports biological production better. Bicarbonate content of water is chiefly for this quality of water, since it acts as a buffer. In events of increasing acidity or alkalinity, the bicarbonate either neutralizes acid by forming calcium of the acid or it binds the hydroxyl radical of the alkali with the precipitation of calcium carbonate as under:

$$Ca(HCO_3)_2 + 2HCl \rightarrow CaCl_2 + 2CO_2 + 2H_2O$$

$$Ca(HCO_3)_2 + KOH \rightarrow CaCO_3 + KHCO_3 + H_2O.$$

Total Hardness of Water

Hardness is the total of soluble Ca (calcium) and Mg (magnesium) salts present in the water, expressed as its $CaCO_3$ equivalent. Total hardness also includes the sulphates and chlorides of Ca and Mg. In most natural waters, the predominant ions are those of bicarbonates, associated mainly with calcium, to a lesser degree with Mg and still less with Na and K. Hardness may be temporary, caused by soluble Ca and Mg bicarbonates, or permanent caused by soluble Ca and Mg carbonates and salts of inorganic acids (*e.g.* $CaSO_2$). Total hardness is used to classify waters as 'hard' or 'soft' but the amount of hardness taken as a criterion for classification varies with different workers. Brige and Juday (1911) offered the following classification:

Soft-water Lakes

Those which contain small amounts of Ca and Mg in solution and in which the average bound CO_2 does not exceed 5 ml per litre.

Medium-class Lakes

Those which contain bound CO_2 between 5 and 22 ml per litre.

Hard-water Lakes

Those which contain more than 22 ml per litre of bound CO_2, which may even exceed 50 ml per litre.

Reid (1966) has categorized lakes into: soft-water ones showing bound CO_2 to be less than 9 or 10 ppm with free CO_2 ranging to nearly 200 ppm and the pH near 6; medium-water lakes characterized by the presence of bound CO_2 up to 30 or 35 ppm, free CO_2 varying widely, frequently showing super saturation, with pH circum-neutral; and hard-water lakes containing bound CO_2 to over 35 or 40 ppm often reaching 200 ppm or more, free CO_2 showing negative value due to withdrawal of bicarbonates at the greater rate than the precipitation of carbonates with pH values lying 8.5 and above. Swingle (1967) has suggested a total hardness of 50 ppm $CaCO_3$ equivalent to be the dividing line between soft and hard waters. According to him, pond waters having a hardness of 15 ppm or above are satisfactory for growth of fish and do not require addition of lime, whereas waters having hardness less than 12 ppm require liming for higher production of fish. Waters less than 5 ppm $CaCO_3$ equivalent cause slow growth, distress and eventual death of fish. According to Swingle (1967), free CO_2 at a concentration of more than 15 ppm is detrimental to pond fishes. The air-breathing fishes can however, survive 100 ppm CO_2.

Dissolved Solids

Both the organic and inorganic solids essential for the growth of fishes are dissolved in water. The inorganic salts are the chlorides, sulphates, carbonates, phosphates and nitrates of calcium, sodium, potassium, magnesium, iron etc. Organic compounds are those of nitrogen and phosphors. Nitrogen compounds are especially significant as they provide food to plants which are consumed by animals. Amino acids, fats, sugar, starch, vitamins are also important organic compounds dissolved in water. Nitrogen is dissolved from atmosphere as well as by denitrification of nitrates, nitrites and ammonia. It is very important for the fertility of a pond. Similarly phosphorus although present in very small quantity in pond water, is important for the production of phytoplankton. It occurs in the form of phosphates. Lack of phosphorus is the main cause of low productivity in fish ponds. Calcium in the form of $CaCO_3$ is highly beneficial, and is utilized in bone building and shell formation. Magnesium is necessary for plants having chlorophyll. Sulphate is also important for the growth of plankton. A number of trace elements like Mn, copper, zinc, aluminium, nickel and cobalt, are present in very minute quantities, but are of grate significance in maintaining productivity of water.

The total concentration of dissolved solids in a water body is a useful parameter in describing the chemical density as a fitness factor and as a general measures of edaphic relationship that contributes to the productivity of the water. Electrical conductivity, which gives the amount of ionized material, is an important measure of total dissolved solids. Electrical conductivity or specific conductance is usually expressed as micro mhos (reciprocal of 1 mega ohms per ml). The electrical conductivity of an electrolyte in solution is directly proportional to the ionic strength of that solution and the total conductivity is the sum of the several conductivities resulting from various ionizable salts present. Sreenivasan (1967) studied electrical conductivity of a number of ponds in relation to fish production and reported that electrical conductivity above 400 m mhos did not limit productivity, but productivity did not increase proportionately with conductivity. The range of electrical conductivity in some rural ponds of Madras has been found to vary from 250 to 3,800 m mhos (Sreenivasan, 1967).

Inorganic Compounds: Nitrogen and its Compounds

Nitrogen is a very important element in pond fertility. Nitrogen occurs in natural waters in elemental (or uncombined) state and as organic as well as inorganic nitrogenous compounds. The elemental nitrogen in water is derived mostly from the atmosphere, the other sources being bacterial denitrification of nitrates, nitrites, and ammonia. The solubility of nitrogen in freshwater varies with temperature and pressure.

Nitrogenous compounds in natural waters may be derived either from outside source (allochthonous) or fixed within the body of the water itself (autochthonous). The former category includes compounds like nitrates, ammonia, etc., carried by rain, surface runoff, inflow of ground waters, seepage, springs, etc. The organic nitrogenous compounds are produced as result of synthesis of inorganic substances into plant and animal tissues. These compounds consist of nitrogen in combination with carbon, hydrogen and oxygen. Organic nitrogen forms about 50 per cent or more of the total soluble nitrogen in surface waters of some lakes, bulk of which is composed of amino-acids, polypeptides and proteins. Many blue-green algae, such as *Anabaena cylindrical, A. gelatinosa, Nostoc muscorum*, etc., secrete extra cellular nitrogenous compounds including polypeptides, amides and amino acids.

Trace Elements

There are certain elements called 'trace elements' or 'minor elements' which, even though occurring in extremely small amounts, are of great significance in imparting productivity to waters.

Manganese

Manganese is commonly found in natural waters. Manganese bicarbonate remains stable in the presence of 1 mg/l of O_2 at *p*H and no oxidation takes place below *p*H 6, as a result of which, dissolved Mn occurs at somewhat higher level of oxygen than iron. In certain tropical lakes, maximum Mn has occasionally been observed dissolved in the upper hypolimnion (metalimnion). The Chaudwar Farm (Orissa) showed manganese values on an average, 0.073 ppm in water and 2.520 ppm in soil. Mn plays an important role in photosynthesis and in heterotrophic growth of phytoplankton. Mn, like iron, is deposited on and in the cell walls of some algae, *e.g.* desmids such as *Penium* and *Closterium* and the thecae of *Trachelomonas*.

Cobalt

Cobalt is present in very small amounts in lake waters (0.02 to 0.09 mg/m3) as a constituent of vitamin B_{12}. This element is required by various blue-green algae. Robbins *et al.* (1950) found in a small pond containing a population of Euglena, a concentration of B_{12} as much as 2 mg/m^3 in late November, but in spring and summer the amount irregularly varied between 0.1 and 0.6 mg/m^3. Das (1967) studied the effect of five micro-nutrients including cobalt on the survival rates of three days old Indian major carp fry (*Catla catla, Labeo rohita* and *Cirrhinus mrigala*) for 11 days. Cobalt gave a survival of 49.75 per cent.

Chapter 8
Physico-chemical Characteristic of Soil

Soil plays an important role in determining the fertility of fish ponds. The basic criterion for selection of a site for construction of ponds is that the soil should not be porous. The soil should be free from vertical and lateral seepage. Barren, slightly sandy, marshy or water logged areas should be chosen as suitable sites. Clayey soil has a very good capacity of retaining water and in this respect such soil is the best of all. Clay is very adsorbent and it tends to hold free water. If the selected site is poor in water retentivity, clay has to be brought from elsewhere and a clay coating and puddle core has to be provided to make the ponds retain water. Large quantities of water and cow dung help in improving the water retention capacity of poor soils.

Soil Fertility

The fertility of soil refers to its nutrient releasing properties for the benefit of land and aquatic crops bearing on it. Soil as a storehouse of nutrients. In the case of aquatic crops, the nutrients are directly required by micro- and macro-vegetation and certain bacteria that may grow in fishery waters.

Pond Mud

Very little attention has been paid towards the study of soils of fish ponds in India. The under water soils of ponds differ from the sub-aerial soils in following respect: (i) they are conglomerated from soils of different profiles completely water-logged and free from air filled spaces, (ii) ponds having a large catchments area receive dissolved nutrients and soil particles through inflowing freshets, and (iii) the sedimentation of organic matter on the pond bottom modifies its properties.

As true pond mud is formed, the productivity of pond increases. The pond mud, which is made of fine soil particles, contains deposits of certain amount of organic matter derived from the bacterial breakdown of plant and animal materials present in water body. Much of the broken down organic matter exists as humus which behaves like an organic compound of higher molecular weight. The humus derived from acid and peaty soils is a mixture of colloidal acids, where as that derived from fertile soils with neutral or weakly alkaline reaction containing lime, comprises calcium salt of the colloidal complex of weak acids. Ponds mud can strongly adsorb phosphate and potash. The pond mud containing humus must undergo alternate processes of reduction and oxidation continuously.

Chemical Control of Soil

Major important chemical factors are *p*H, total nitrogen, total phosphorus, organic carbon, C/N ratio, available nitrogen, phosphorus, potassium and exchangeable calcium.

Hydrogen Ion Concentration

The pH of the soil depends on various factors. Decomposing of organic matter formed hydrogen sulphide, methane, and short chain of fatty acids which make the soil acidic, it reduces the rate of bacterial action influencing productivity.

The pH of the soil also influences transformation of soluble phosphates and controls the adsorption and release of ions of essential nutrients at soil-water interface. Near neutral pH 7.5–8.5 is suitable. pH control by liming the soil.

Phosphorus

Phosphorus occurs in soil in both inorganic and organic forms. The inorganic phosphates in the soil can be classified into four main groups: calcium phosphate, aluminium phosphate, iron phosphate, and the reductant soluble phosphate extractable after removal of the first three forms. The organic phosphorus compounds in soils are known to occur in three main groups: (a) phytin and phytin derivatives, (b) nucleic acids, and (c) phospholipids.

It is the availability and not amount of phosphorus which is more important to aquatic productivity owing to the fact that PO_4 ions in soil form insoluble compounds with iron and aluminium under acidic conditions and with calcium under alkaline conditions, rendering the phosphorus ion unavailable to water phase under either condition. A considerable amount of phosphatic ions also remains adsorbed on colloidal complexes. Available soil phosphorus shows a definite positive correlation with fish production. Banerjea (1967) has grouped pond soils studied by him under four ranges of available phosphorus: < 30 ppm, 30 to 60 ppm, 60 to 120 ppm and >120 ppm. Soil phosphorus (P_2O_5) level <30 ppm is considered indicative of poor, the range 30 to 60 ppm of average, and the level > 60 ppm of high, fish production.

Calcium Bound Phosphate

Calcium bound phosphate fraction shows the highest occurrence in alkaline soils. When phosphate fertilizers are added to pond waters, the uptake of phosphorus by phytoplankton takes place very quickly. The ability of the phytoplankton to rapidly absorb phosphate added to fish ponds. Alkaline soil had adsorbed more phosphorus than acidic.

Calcium-Carbonate Phosphate System

Hepher (1958) mentioned that in ponds having alkaline water and high Ca content, most of phosphorus added is precipitated as $Ca_3(PO_4)_2$. Decomposition of organic colloids regenerates phosphorus which counter acts their adsorptive roles and CO_2 is also evolved which lowers the *p*H of mud, there by reducing the fixation of phosphorus by $CaCO_3$. Phosphorus fixation is higher when the mud is mixed with the water of the pond. In fish ponds this is done by some fishes searching food in the bottom mud.

Nitrogen

The role of nitrogen in soils of fish ponds has been reviewed by Neess (1949) and Hepher (1952). The organic matter occurring on pond bottom is broken down by a number of microorganisms. The end products of this decomposition among others are CO_2, water and ammonia. While ammonium salts are formed, a number of bacteria found in the soil and water convert ammonia nitrogen into nitrates. Further changes of ammonium salts and nitrates occur by the following processes.

(*a*) Denitrification

Nitrates and ammonia salts are reduced to free nitrogen through the agency of denitrifying bacteria. The denitrification depends on conditions prevailing in the pond, low concentration of dissolved oxygen in the water accelerating their activity and thereby increasing the release of nitrogen (Hepher, 1952). Nitrate acts as a hydrogen acceptor alternatively to the oxygen. Golterman (1967) reported that in fish ponds, where a high NO_3 concentration is desirable, while the pond probably has a high oxygen consumption, the nitrate is easily reduced to ammonia.

(*b*) Leaching

Ammonia salts and nitrates are highly soluble compounds and are, therefore, easily leached out with the percolating water.

(*c*) Volatilization of Gaseous Ammonia

A part of the ammonia released during the bacterial decomposition of organic matter can also become volatile, especially when the temperature is high and the water is alkaline. This is also influenced by the amount of lime added to the pond. The factors which cause nitrogen supply to the pond on the one hand, and its escape from the pond on the other, create an equilibrium between the nitrogen contents of water, soil and air. Both nitrogen and phosphorus become limiting factors and should be added to the pond in the right proportion, so that at least a part of them is utilized by the plankton in the pond. The soil is the main source of nitrogen supply to the pond through the decomposition of organic matter. The application of nitrogen fertilizers causes a decrease in the activity of nitrogen fixing bacteria and at the same time stimulates that of the denitrifying bacteria. Banerjea (1967) observed fish production was poor in ponds with nitrogen below 25 (the unit being expressed as mg N/100 g soil. In the ranges, 25 to 50 and 50 to 75, the production was average.

Organic Carbon and C/N Ratio

The significance of organic fertilizers lies in their carbohydrate content. Bacterial activity depends not only on the carbon content but also on the C/N ratio. Nitrogen fixation by Azotobacter must proceed in a substrate of carbohydrate materials, required both as a source of energy and as a constituent of bacterial protoplasm. Carbon from carbohydrate is taken up with free nitrogen in certain definite ratios. The fixation stops if readily useful carbohydrates are not available. The bacterial activity is low when C/N ratio falls below 10:1 and high when the ratio is 20:1 or higher (Meehean, 1935). Banerjea (1967) has studied organic carbon content and C/N ratio in a number of ponds. The range of 0.5 to 1.5 per cent of organic carbon showed average fish production while 1.5 to 2.5 per cent seemed to be optimal Though relatively better production is found in the ratio range 5 to 10, the range 10 to 15 represents the ideal condition.

Table 8.1: Relation Between Soil Condition and Productivity of Fish Ponds (Banerjea 1967)

Parameter (Range)	*No. of Ponds*			
	Total	*Productive*	*Average*	*Unproductive*
		pH		
5.5 (Highly acidic)	2	-	-	2
5.5-6.5 (Moderately acidic)	22	2	9	11
6.5-7.5 (Near neutral)	25	13	4	8
7.5-8.5 (Moderately alkaline)	25	7	11	7
8.5 (Highly alkaline)	6	-	3	3
		Available N (mg/100g)		
25	39	4	21	14
25-50	27	10	6	11
50-75	11	6	-	5
75	3	1	1	1
		Available P_2O_5 (mg/100g)		
3	28	-	-	28
3-6	27	-	26	1
6-12	18	14	3	1
12	7	7	-	-

Contd...

Table 8.1–Contd..

Parameter (Range)	No. of Ponds			
	Total	Productive	Average	Unproductive
Organic Carbon (%)				
5	20	1	8	11
0.5-1.5	37	6	19	12
1.5-2.5	17	14	-	3
2.5	6	1	2	3
C/N Ratio				
5	20	1	8	11
5-10	37	6	19	12
10-15	17	14	-	3
15	6	1	2	3
Exchangeable Calcium (mg/100g)				
100	17	4	4	9
100-200	18	7	6	5
200-300	18	7	4	7
300	27	3	15	9

Potassium

Potassium is taken up readily by submerged weeds for growth, especially *Potamogeton*, *Myriophyllum*, *Ceratophyllum*, etc. During rapid plant growth period, potassium from the water and soil is stored in the tissues. Ponds with sandy, non-rapid adsorptive soils are usually potassium-poor and respond most markedly to fertilization. Potassium is adsorbed to the mud during winter and released during summer. Worbel (1967) reported that exchange of potassium between soil and water is relatively less dependent on the carbon turn-over cycle. According to him, insoluble potassium salts are not formed in ponds. For exchange of potassium from soil to water, a high hydration of soil colloids is favorable, and for that reason, there is never a total absence of this nutrient in ponds. Only in ponds located on peat bogs, or those infested with submerged weeds, potassium content may fall below 0.5 mg/l.

Calcium

Calcium is considered important owing to its four main roles: (i) its relation to translocation of carbohydrates, (ii) as an integral component of plant tissue, (iii) for increasing the availability of other ions, and (iv) for reducing the toxic effect of single salt solution of other elements (Welch, 1952). In pond soil, calcium is generally present as $CaCO_3$. According to Barret (1953), the amount of exchangeable phosphate in pond mud is inversely related to the marl-organic matter ratio, so that in highly organic soil with low calcium, soluble phosphorus remains adsorbed in an exchangeable form but, in sediments having low organic matter and high quantity of marl, phosphorus probably becomes fixed in the form of insoluble precipitate.

Chapter 9
Pond Fertilization

The purpose of fertilization is to increase the productivity by increasing the natural food available. The action of the fertilizer is rather indirect. The fertilizing substances in the fertilizer provide the necessary nutrient substances, which are either lacking in the pond water or are insufficient. The supply, supplement or increase of these nutrient substances, is responsible for increase of these nutrient substances, is responsible for increased production of producer of the first tropic level, the plants (the phytoplankton and vegetation). The addition of the fertilizers to pond, is immediately absorbed by the bottom mud which later releases the nutrients in minimal quantities into the water for long periods.

For best results, the pond to be fertilized should be prepared before hand as follows:

1. The acidity of the pond (soil and water) should have been corrected so as to have a neutral or slightly alkaline conditions.
2. All emergent weeds should have been destroyed and removed (for they are the worst indirect competitors for fish).
3. The floating and submerged weeds should have been properly controlled so that sunlight penetration is not interfered with.

4. The pond bottom should have a good mud (colloid of fine detritus) but not a mud of cellulose (produced by dead and decomposing emergent weeds).

The fertilizers are classified into two categories:

1. Inorganic fertilizers or mineral fertilizers, and
2. Organic fertilizers or manures of plant and animal origin.

The chief inorganic fertilizers include lime, phosphate, potash, nitrogenous compounds and magnesium fertilizers. Inorganic fertilizers causes phytoplankton blooms, and heavy growth of zooplanktons. Organic fertilizers comprise green manures the compost and the manures of the animal origin such as liquid from cattle farms: sewage, dung etc. Organic fertilizers carry almost all the nutrients required by the fish. The liquid manure from stables is very beneficial and stimulates the growth of planktons, as well as filamentous algae. Farm manures consisting of cow and pig dung, is extensively used in fish culture. Sewage is also considered a good fertilizer with nitrogen and phosphorus as its main constituents. Organic fertilizers increase the fish production as they provide almost all the nutrients needed in their biological cycle. Other substances used as manures are soybean meal, cotton seed, mustard oil cake alone or in combination with cow dung is extensively used in India for pond fertilization. Proportionate quantity of organic manure in the form of dry cow dung for nursery pond is to be applied. Overdoses may have adverse effects on the growth of fish fry, particularly in causing the gill rot and oxygen depletion. Before application, the organic manure is made into a paste with water and applied at different parts of the pond. To ascertain uniform distribution of organic manure, the pond is dragged using the net.

Fertilizers

The fertilizer used in fish ponds fall under two categories:

1. Inorganic, and
2. Organic

Inorganic Fertilizers

The major fertilizing elements are nitrogen, phosphorus, potassium and calcium. Minor elements having a manurial value are manganese, boron, sulphur, iron, copper and zinc. The soluble inorganic fertilizers make their elements available to the water immediately they are applied. The inorganic fertilizers are customarily expressed as percentage of available nitrogen (N), phosphoric acid (P_2O_5) and potash (K_2O). The different inorganic fertilizers used in fish pond are grouped as:

1. Limestone and lime containing fertilizers;
2. Phosphatic fertilizers;
3. Nitrogen fertilizers;
4. Potassium fertilizers;
5. Magnesium fertilizers; and
6. Trace element fertilizers.

Organic Manures

Organic manures, as a class, are composite and contain almost all the nutrient elements required in the metabolic cycle. They enrich the organic matter content of soil and water and, within the limits of manurial dose, release carbon dioxide and nutrients on decomposition, sustaining the fertility of water. Their merit lies in imparting a comparatively slower rate of release of nutrient to the water over a longer period of time as the decomposition proceeds. Several kinds of useful bacteria may also get incidentally introduced along with the organic fertilizers in to the ecosystem.

The use of organic manures in fishery water, already having thick organic sediment at the bottom, hastens the depletion of dissolved oxygen and enhances production of toxic gases. Manures of organic origin have been classified by Mortimer and Hickling (1958) as follows:

1. Those containing little or no carbohydrate, *e.g.* dried blood, urine, guano, offal and liquid manure;
2. Those containing considerable carbohydrate with nitrogenous matter, *e.g.* farmyard manure, brewery refuse, sewage, sludge etc;
3. Manures containing mainly carbohydrate, *e.g.* plant material, cellulose waste, wood shavings, molasses, cotton-seed meal, etc.

A detailed description of the inorganic and organic manures follows.

Inorganic Fertilizers

(*a*) Lime Stone and Fertilizers Containing Lime

The first step in fertilization is the application of lime which is available in several forms, the commonest being lime stone. Lime has been considered as a fertilizer owing to the fact that it supplies calcium, one of the essential nutrients (Hickling, 1962). The chief uses of liming are:

(*i*) It corrects the acidity of soil and water; (*ii*) The above process, possibly aided by the presence of calcium, speed up the decomposition of organic matter, releasing carbon dioxide from the bottom sediments; (*iii*) It raise the bicarbonate content and lack of CO_2 does not become a limiting factor; (*iv*) Liming a pond establishes a strong *p*H buffer system. According to Wolny (1967), this system comprises CO_2-HCO_3-$CACO_3$ preventing wide fluctuation in the *p*H of water; (*v*) It counteracts the poisonous effect of excess magnesium, potassium and sodium ions; (*vi*) It fixes harmful acids (such as humic acid) and inorganic acid (such as sulphuric acid); (*vii*) The calcium content of lime displaces certain other fertilizers from organic colloidal system, thus making available grater quantities of such materials as K^+ and PO_4^-, when they are applied as manure in the pond; (*viii*) Lime, by its toxic and caustic action, kills bacteria as well as fish parasites in their various life history stages, and may render the fishes less liable to disease.

According to Hora and Pillay (1962), large perennial waters are to be slightly acidic, liming @ 300 to 500 kg/ha. Further, quicklime should be used @ 200 kg/ha to control the gill rot disease, but quantity should be regulated to maintain *p*H below 9. For a through disinfections of ponds, a dose of about 10,000 kg/ha is needed, but if liming is regularly done every year for fertilizing purposes, 100 to 200 kg/ha suffice for soils which are not very acid or poor in carbonates. A dressing of 1,000 to 1,500 kg/ha is required for ponds with soil and water. It is concluded that liming is an essential preliminary to successful pond manuring. The dose lime depends upon the characteristics of the soil and water.

The advantages of using quicklime are: (*i*) quicklime, being almost pure and containing 98 per cent CaO, is less expensive to transport than limestone, which contains 50 per cent of CaO only; (*ii*) it can bring much greater quantities of CO_2 compared to capacity of limestone; (*iii*) quicklime takes more time than limestone to take up CaO_2 from the air or soil necessitating that ponds be not stocked with fish until at least a fortnight after treating with quicklime or caustic lime; (*iv*) it has greater capacity than $CaCO_3$ to precipitate organic matter and clear the water: and (*v*) it is a powerful disinfectant and fish poison.

Lime can be applied to the pond bottom, added to the water at the inlet, or spread on the water surface. Slaked lime and quicklime should, as far as possible, be applied to ponds after draining.

Table 9.1: Lime Doses at Different *p*H Values of Soils

pH	*Soil Condition*	*Dose of Lime (kg/ha)*
4.0–4.5	Highly acidic	1000
4.5–5.5	Medium acidic	700
5.5–6.5	Slight acidic	500
6.5–7.5	Near neutral	200

(*b*) Phosphatic Fertilizers

Phosphorus, though required in small quantities, is the most important single element controlling organic production in natural waters. Lin (1970) reported that calcium superphosphate @ 2,000 kg/ha gave the best results in boosting production of fish in freshwater ponds in Taiwan, especially that of silver carp.

Saha and Chatterjee (1979) made a comparative study on the availability of phosphorus from phosphatic fertilizers in relation to pond productivity and found triple superphosphate (40 per cent P_2O_5) to be more efficient in maintaining an higher level of soluble phosphorus in the water phase while single super phosphate (16 per cent P_2O_5) maintained a higher level of available phosphorus in the soil phase. Triple superphosphate at 25 kg/ha when applied to slightly acidic soil (*p*H 6.0) gave the best results in terms of survival and growth of rohu spawn.

Nitrogen manures are reported to be effective (Demoll, 1925) only when there is a definite nitrogen deficiency in pond. The important inorganic nitrogenous fertilizers used in fish ponds to supply nitrogen are described below:

(*i*) Chile Saltpeter (Sodium Nitrate)

Chile saltpeter has about 16 per cent nitrogen and contains, as impurities, a number of elements like potassium, boron, magnesium, iodine, calcium, chlorine and sulphar. It is readily soluble in water. Kuhnert (1910) observed a marked increase in fish yield in ponds fertilized with Chile saltpeter along with sulphate and potassium salts.

(*ii*) Ammonium Sulphate

Ammonium sulphate contains 20 per cent nitrogen and is one of the most widely used fertilizers throughout the world. Ponds fertilized with ammonium sulphate promptly give rise to phytoplankton blooms. The fertilizer has also been observed to stimulate the growth of bottom fauna and lead to sustained production of zooplankton over the season (Schaperclaus, 1966). Bank (1959) demonstrates that fish grow better in ponds fertilized with ammonium sulphate.

Yard trials with 80 kg N/ha of ammonium sulphate, when applied to slightly acid, nutrient-deficient soil, gave 39 per cent survival of major carp spawn. A very encouraging result in spawn survival was achieved in a nursery pond (soil *p*H 6.9) treated with only urea (80 kg N/ha) when, without artificial feeding, survival as high as 76 per cent of rohu spawn.

(*iii*) Ammonium Nitrate (Norway Saltpeter)

It contains 20 per cent pure nitrogen in the ammonium form and 15 per cent nitrogen in the nitrate form. It stimulates the growth of phytoplankton. Wolny (1967), the fertilizers (super phosphate 400 kg/ha and Norway saltpeter 800 kg/ha) were administered to the water in small lots throughout the growing season. The weight increment of fish came to 400 kg/ha. The mean consumption of fertilizers per 1 kg carp spawn growth was 0.67 kg N and 0.25 kg P_2O_5.

(iv) Ammonia Liquor

Ammonia liquor containing 20 per cent nitrogen. Hepher (1959) found that aqueous ammonia as just as effective as a fertilizer like sulphate of ammonia, when ammonia is used, it raises immediately *p*H of the pond. The rate of application was 56 kg per dunam (1/10 ha) of ammonium sulphate and 45.5 kg per dunam of 20 per cent aqueous ammonia.

(v) Urea

Urea contain 45 to 46 per cent pure nitrogen. It is a bicomponent fertilizer with 78 kg CO_2/100 kg of urea. Wrobel (1962) considers it to be a slow action fertilizer because, in this investigations, phytoplankton first appeared in the latter part of the growing season.

(c) Potassium Fertilizers

In India, preliminary experiments at the pond Culture Substation of CIFRI have shown that the application of 18: 8: 4:: N: P: K @ 500 kg/ha gave good results in the preparation of nursery ponds and also in short-term rearing experiments. Pond soils to 18: 8: 4:: N: P: K mixture using it @ 90-40-20 (N-P_2O_5-K_2O) kg/ha, in one installment and the same amount in three split up doses. Moderately alkaline soil with very low available phosphorus showed the maximum response to the fertilizer, increased production being 472.0 and 363.2 kg/ha/annum for single and divided dose of fertilizers respectively.

(d) Magnesium Fertilizers

According to the Hickling (1962), magnesium may prove beneficial in fish ponds, where the soil or land drainage is deficient in the element or in the element or in the situations in which the level of other essential nutrients has been raised to a point where magnesium itself may become a factor limiting further increased production. Magnesium fertilizers may be useful in peaty mud.

(e) Trace Element Fertilizers

Not much is known about the role of trace elements like zinc, manganese, iron, copper, boron, molybdenum, iodine, etc., in fish culture. Sreenivasan *et al.* (1975) studied the effects of zinc, molybdenum, cobalt, boron, chromium, barium, selenium and stannous tin at 0.05 ppm and that of zinc, molybdenum, cobalt and boron at 0.05 ppm resulted in the highest plankton volume and primary production. Among the elements tried at 0.05 also, zinc gave the highest primary production values followed by chromium, molybdenum, boron, barium, cobalt, selenium and stannous tin.

Organic Manures

(*a*) Organic Manures Containing Little or No Carbohydrates

Liquid Manure

The liquid manure, from stables and byres is a beneficial fertilizer which stimulates the growth of zooplankton and phytoplankton, filamentous algae and hard water plants. Ponds treated with liquid manure give high fish yield. Liquid manure should be applied in repeated small doses in deeper parts of the pond so as to avoid its wastage on marginal vegetation. The use of guano @ 3 dz/ha (1 dz per ha = 89.2 1b per acre) has been reported to give fish increase from 100-150 per cent (Schaperclaus, 1933).

(*b*) Organic Manures Containing Considerable Carbohydrate with Nitrogenous Matter

Farmyard Manure

The term farmyard or farm manure is employed to denote refuse from all animals of the farm. Farm manure consists of two original components-the solid and the liquid-in approximate proportion of 3:1. Cow and pig dung are very frequently applied. These animal manures, especially liquid and farm manures, are very useful for conditioning the soil of a new pond, and providing a ready-made mass of organic matter containing the necessary nutrients. The organic matter helps in the formation of a good pond mud, promotes nitrification and formation of a good colloidal structure.

In India, Alikunhi *et al.* (1955) manured the nursery ponds of the Cuttack Substation of CIFRI with cow dung at the rate of *c.* 11,208 to 22,417 kg per ha. Sharma *et al.* (1979) observed that droppings from 100 to 150 birds (10 to 15 t/year), fed on concentrated poultry feed were sufficient to fertilize 1 ha water area. A sustained high plankton production was obtained in the pond resulting in fish production of the 4.3 t/ha/yr.

Dung from camel and horse stables is also used in several countries. Night-soil is used for manuring ponds in Hong Kong and China. In an experiment, in Hong Kong, 6,028 kg of raw night-soil and 628 kg of soyabean waste were applied per ha per month to obtain a fish yield of *c.* 1,680 kg/ha/annum (Hickling, 1962).

Sewage is considered to a rich fertilizer with nitrogen and phosphorus as its main chemical constituents. It is a common practice in various countries to use only diluted sewage as a pond fertilizer. In Calcutta, sewage is admitted slowly in certain ponds periodically in the proportion of 1:4 sewage to water.

Sludge is used as pond fertilizer in many countries. In India, Gopalakrishnan and Srinath (1963) used activated sludge as a manure in cement cisterns was found to increase the growth rate of fish and enhance the production of plankton through release of nutrients into water. However, digested sewage sludge cake it contains 3.84. to 4.79 per cent nitrogen and 1.55 to 1.20 per cent phosphorus. Natarajan and Vorghese (1979) found that, digested sludge cake was used at 2,250 kg/ha in a single dose or at 3,495 kg/ha in two split doses. Significantly faster growth rates of catla and common carp were recorded in the fertilized cisterns when compared with those in the control.

(*c*) Organic Manures Containing Mainly Carbohydrates

Soybean meal, cotton-seed meal, tea-seed cake, mustard oilcake, mahua oilcake, etc., are commonly used for manuring fish ponds. Smith and Swingle (1942) obtained increased fish production in ponds fertilized with cotton-seed meal, soybean meal, and poultry mash.

In India, oilcake (mainly mustard oilcake) is used singly or in combination with cow dung, for manuring fish ponds. Cuttack Substation of the CIFRI with a view to evolving a suitable method for fertilizing nursery ponds, with mixed inorganic fertilizer 6:8:4:: N:P:K and organic fertilizer, mustard oil cake, applied singly and in combination on equivalent nitrogen basis @ 12 kg/ha/annum. Best results were obtained with N-P-K + oil cake (Barrackpore, 1963). Mahua oilcake, which acts as a poison for, killing fish in the initial stage of application, serves as a fertilizer subsequently leading to development of plankton. Plankton production with groundnut oil cake, horse dung and poultry droppings was significantly higher than inorganic fertilizer.

Green manures refers to turning into the soil undecomposed green plant tissue which is often done by sowing a nitrogenous or other crop on dry pond bottom. The advantages of green manures are: (*i*) it supplies organic matter, which, besides acting as food for soil organisms, induces microbial activity in the soil raising its productivity, (*ii*) the humus compounds formed on the decay of green manure increase the adsorptive capacity of the soil promoting its aeration, drainage and granulation, (*iii*) it restores soil nitrogen if leguminous plants are used as green manure, (*iv*) the green manuring crop exerts conserving influence on the nutrient of soil science it take up soluble constituents that might be lost by drainage or erosion, (*v*) when the crop is ploughed in, the nutrient carried by the roots from below, are left near the surface, (*vi*) green manuring is believed to increase the availability and solubility of lime and phosphoric acid in soil and, after decomposition, provides inorganic nutrients directly to the soil at the same time influencing potassium, magnesium and iron contents, and (*vii*) it promotes a suitable substratum in the pond for a rich growth of attached algae, zooplanktons, insect larvae, worms, etc., which form the fish food.

In India, green manure when applied, is either ploughed into the soil at the time of preparing ponds or dumped in heaps thereon. Hora (1945a) stated that in Bengal green grass along with dry cow dung, stable refuse, poultry manure and oil cake is used for manuring fish ponds.

Compost is commonly made by dumping alternately into a large pit, layers of manure and vegetable matter. Layers of sod or of soil, high in organic matter, may also be introduced in the composting pits. The manure supplies especially active anaerobic bacteria which, in the presence of nitrogen, cause rapid and effective humification. This results in reducing the percentage of carbon and increasing that of nitrogen. The compost pit is preferably covered with earth. Hora and Pillay (1962) stated that any available plant matter, such as leaves, cut grass, aquatic vegetation, etc., can be composted by dumping them in layers about 30 cm high alternately with 7.5 cm high layers of manure dusted in between with superphosphate of lime. A 1:10 ratio of nitrogen and 1000 kg of compost and, in order to maintain the necessary humidity, *c.* 4,000 l of water is also added. The compost, if upturned, after intervals of 5 and 8 weeks, should be ready for use in about 3 months. Hora and Pillay (1962) suggested a dose of 5,000 kg of compost per ha in ponds which will give rise to a rich growth of plankton. In India, various combinations of cow dung, mustard oilcake and aquatic weeds such as *Eichhornia, Hydrilla, Najas*, etc., have been suggested for composting and manuring at the rate of approximately 555 to 832 kg/ha (Saha, 1953).

Fertilization of Nursery Pond

Natural food production in the nursery ponds is limited and when a large number of fry are stocked the available food is not sufficient for the growth. The object of manuring the nursery ponds is to the production of zooplanktons which form the natural food of carp fry. After removal of predatory and weed fishes the first step is liming the pond, the dose depending upon the *p*H of the soil.

The recommended application rate is as follows:

pH of Mud	*Limestone kg/ha*
Less than 4	6000–12000
4.0–4.5	4000–9600
4.5–5.0	3600–7200
5.0–5.5	3000–4800
5.5–6.0	1600–3000
6.0–6.5	1400–1600

After liming the ponds are treated either with organic manures such as cowdung or inorganic fertilizers (N.P.K). Chemical fertilizers like the superphosphates, ammonium sulphate and sodium nitrate may be used but are costly. Moreover, they bring about growth of microscopic algae which are not used as food by carp fry.

Hence, cow dung as an organic manure is preferred. The dose depends on various factors including the fish poison used. For example, if mahua-oil cake has been used as a fish poison, which is manure also, only 5,000 kg of cow dung per ha is enough.

Generally 5,000-10,000 kg of cow dung/ha brings about thick swarms of zooplanktons like rotifers, copepods, and cladocera, is about 15-20 days. Manuring is done 10 days after poisoning. The cow dung can be spread on the pond bottom before letting in water, or spread on the surface or deposited in heaps along with banks. It is advisable to apply at suitable intervals in two doses instead of a single dose. Thus, an initial dose of 10,000 kg of cow dung/ha, two weeks before stocking the ponds, followed by a second dose of 5,000 kg/ha, a week after stocking is reported to maintain sustained production of zooplanktons for a long period.

Recently, a mixture of superphosphate of lime, cow dung and mahua–oil cake has been used successfully for manuring. The mixture is converted into a thick paste with water and spread over the pond surface.

Fertilization of Rearing Pond

The spawn stocked in nursery ponds ordinarily attain a length of 20-25 mm in 15 days with artificial feeding. Then the fry are netted out and released in rearing pond, which are to be prepared more or less on the same lines as the nursery ponds. The management of rearing ponds is aimed at culturing maximum number of fry to fingerlings within shortest period of time.

Cow dung 11,230 kg/ha 10 days before stocking and ammonium sulphate + single superphosphate + calcium ammonium nitrate (11:5:1) at 690 kg/ha, 2 months after stocking.

Fertilization of Stocking Pond

Continued cultivation of fish in a pond will steadily decrease the crop. To maintain and augment production the pond will have, therefore, to be manured.

Carp fry which commonly attain a size range of 75 to 125 mm within 2 months in rearing ponds are netted out and released in stocking ponds. Stocking pond are large perennial water bodies varying from 0.2 to 2.0 ha in area with an average depth of 1.5 to 2 m.

The goal of the management of stocking ponds is to attain high production of table sized fish in the shortest possible time. The first step requisite for the stocking ponds is to clear them of weeds and remove predatory and weed fishes the methods same as used for nursery pond. In order to augment production of natural food fish organisms in small stocking ponds it is necessary to manure them in suitable installments with organic manure. Cow dung 20,000 to

25,000 kg/ha/yr and inorganic fertilizers mixture of ammonium sulphate + single superphosphate + calcium ammonium nitrate in 11:5:1 ratio at 1,000 to 1,500 kg/ha/yr, which may suitably varies depending on the water and soil quality.

Chapter 10

Predatory and Weed Fishes and their Control

The term "weed" fish is used here to include all species of uneconomic, small sized fish that naturally occur or are accidentally introduced in ponds along with carp spawn. Weed fishes have high fecundity and ripen sexually in summer months. They breed in ponds, even in the absence of rains and their young abound in ponds during the monsoon period. Predatory fishes are harmful to cultured major carps as not only they compete with the latter for food and space but also prey on them. Normally, some of the predatory fishes breed in ponds a little earlier than the major carps spawn in riverine habitat or immediately at the onset of monsoon and their young ones feed vigorously on the plankton available in pond and grow so fish that when the carp spawn is released in to nursery, the predators are large enough to take a heavy toll of the spawn. Predatory fishes are larger than the young ones of major craps in the same age group. Most of the predatory fishes are also food fishes.

The commonly occurring predatory and weed fishes in nursery ponds are given below:

Predatory Fishes

Channa marulius	*Channa punctatus*
Channa striatus	*Channa gachua*
Channa stewarti	*Anabas testudineus*

Nandus nandus	*Glossogobius giuris*
Notopterus notopterus	*Notopterus chitala*
Heteropneustes fossilis	*Clarias batrachus*
Lates caclarifer	*Wallago attu*
Ompok pabo	*Ompok pabada*
Ompok bimaculatus	*Silonia silondia*
Ailia coila	*Clupisoma garua*
Eutropiichthys vacha	*Gagata cenia*
Glyptothorax sp.	*Mystus seenghala*
Mystus aor	*Mystus cavasius*
Mystus bleekeri	*Pama pama*
Pseudosciaena coitor	*Mastocembelus pancalus*
Mastocembelus armatus	*Macrognathus aculeeates*
Anguilla bengalensis	*Amphipnous cuchia; Garra* sp

Weed Fishes

Ambassis ranga	*Ambassis nama*
Amblpharyngodon mola	*Aspidoparia morar*
Barilius barila	*Barilius bendelisis*
Barilius bola	*Barilius vagra*
Chagunius chagunia	*Chela laubuca*
Chela gora	*Esomus danrius*
Osteobrama cotio	*Oxygaster bacaila*
Oxygaoter phulo	*Puntius conchonius*
Puntius phutinio	*Puntius sarana*
Puntius sophore	*Puntius ticto*
Gudusia chapra	*Gonialosa monmina*
Setipinna phasa	*Rhinomugil corsula*
Sicamugil cacasia	*Xenentodon cancila*
Colisa sp	*Botia dayi; Noemacheilus zonahis*

Figure 10.1: *Mystus cavasius*

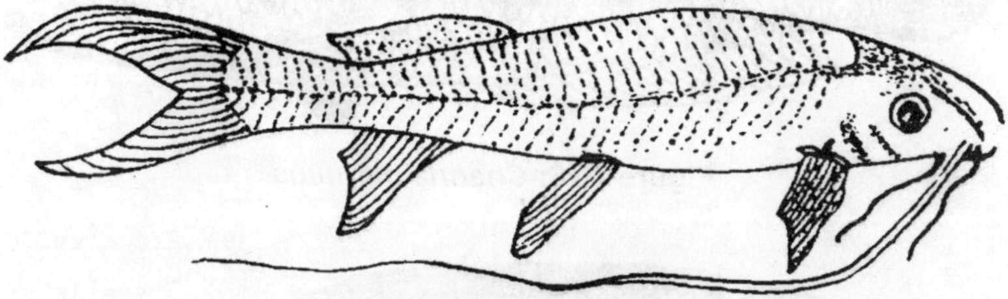

Figure 10.2: *Mystus aor*

Figure 10.3: *Heteropneustes fossilis*

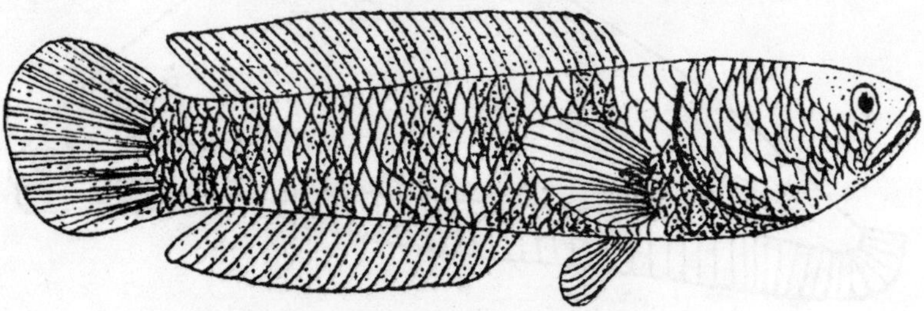

Figure 10.4: *Channa punctatus*

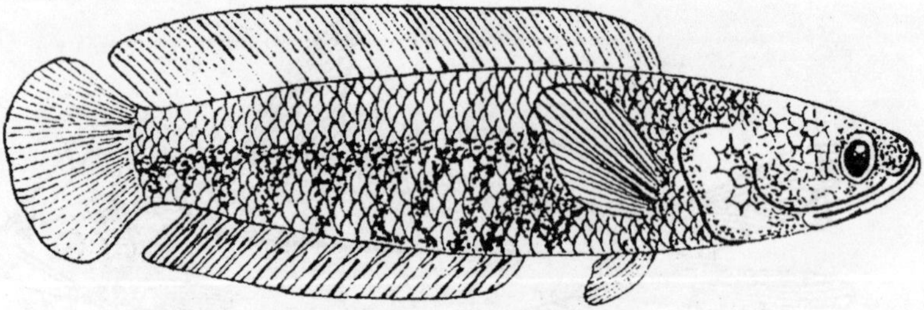

Figure 10.5: *Channa striatus*

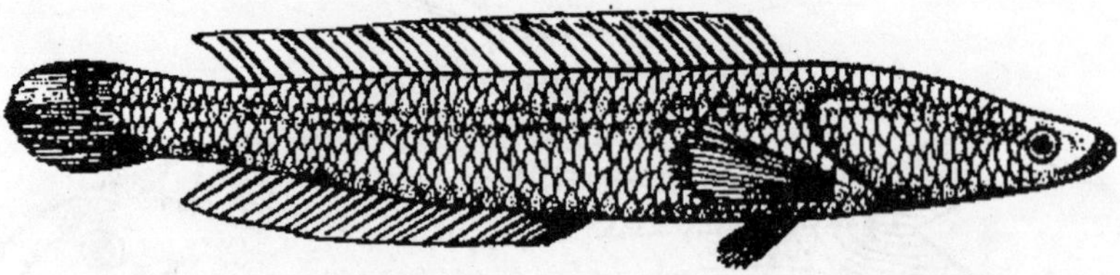

Figure 10.6: *Channa marulius*

Figure 10.7: *Clarias batrachus*

Figure 10.8: *Notopterus chitala*

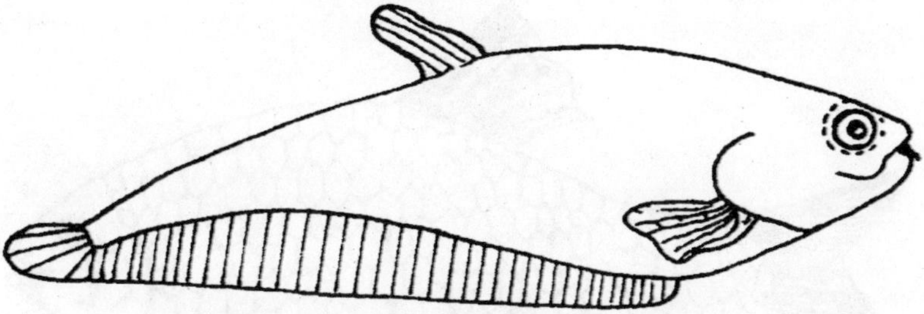

Figure 10.9: *Notopterus notopterus*

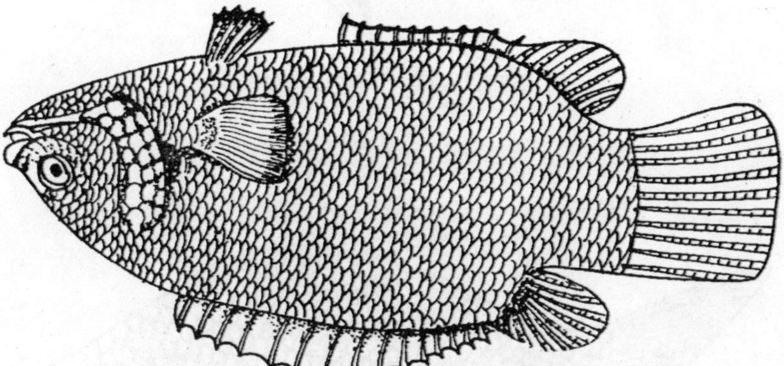

Figure 10.10: *Anabas testudineus*

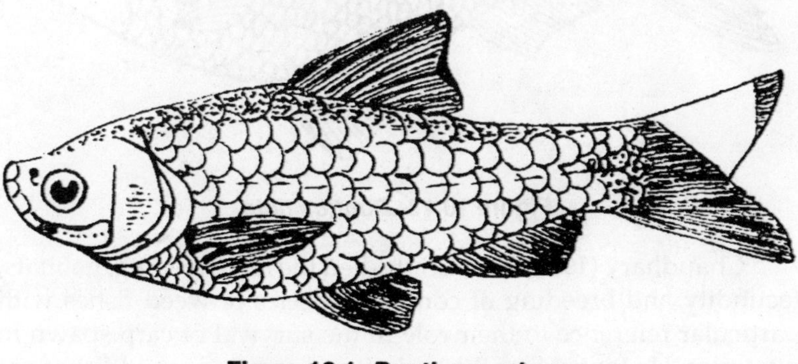

Figure 10.1: *Puntius sophore*

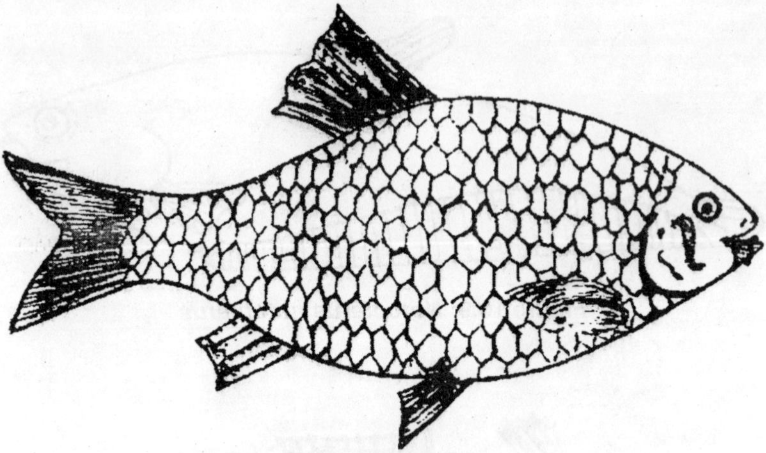

Figure 10.12: ***Puntius sarana***

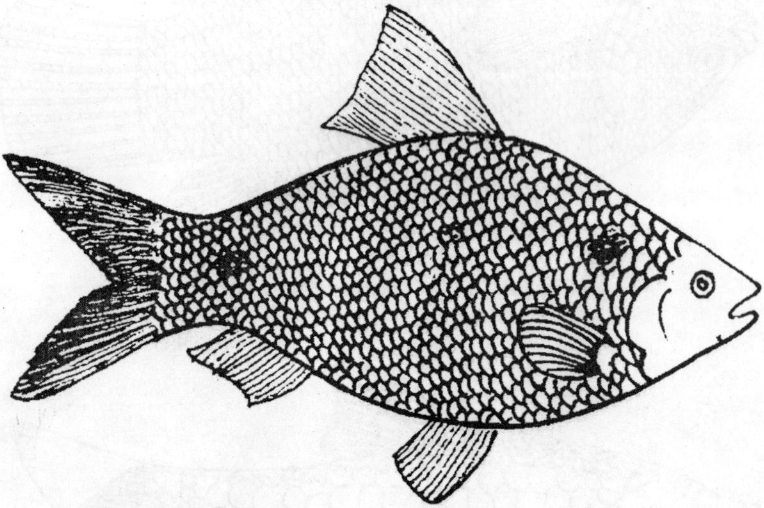

Figure 10.13: ***Puntius ticto***

Chaudhary (1960) has also studied the food, feeding, habitats, fecundity and breeding of common species of weed fishes with particular reference to their role in the survival of carp spawn in nurseries. He found that the majority of common weed fishes are active competitions with the major carps for the food available in pond. Consume large quantities of zooplanktons, the main food of

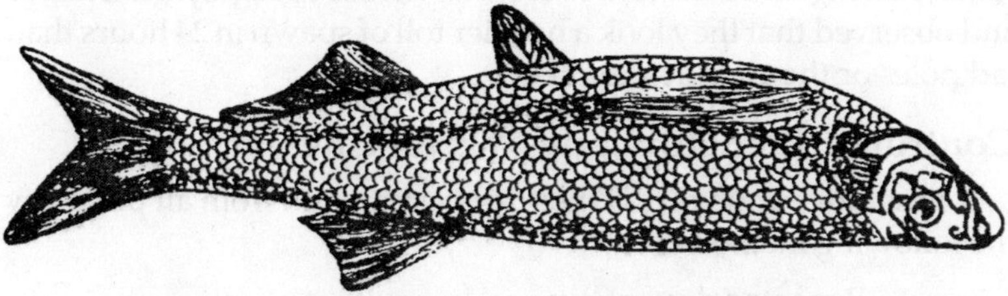

Figure 10.14: *Oxygaster bacaila*

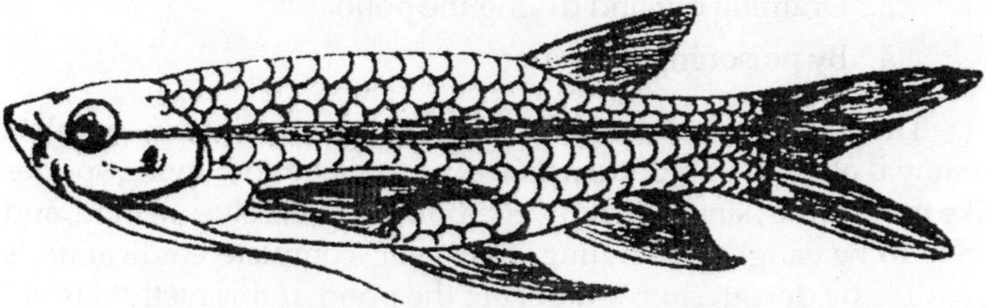

Figure 10.15: *Esomus danricus*

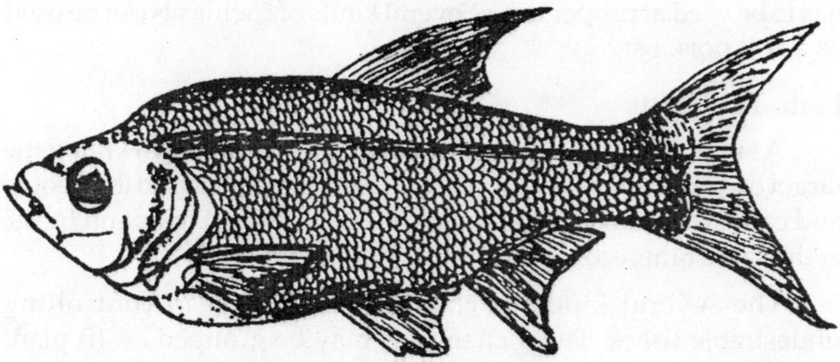

Figure 10.16: *Amblpharyngodon mola*

the carp spawn. Weed fishes have high fecundity and ripen sexually in summer months. They breed in ponds, even in the absence of rains and their young abound in ponds during the monsoon period. According to Alikunhi *et al.* (1952), even the young weed fishes directly feed on carp hatchlings and spawn and take a significant toll there of as soon as they are released in nurseries. Tripathi and Sharaf (1974) listed *Puntius sophar* followed by *Chela laubuca*,

Amblypharengodon mola and *Rasbora danconius* as highly destructive and observed that they look a heavier toll of spawn in 24 hours than tad poles or the aquatic insects.

Control of Predatory and Weed Fishes

Predatory and weed fishes can be removed from all ponds by the following methods:

1. Repeated drag netting
2. Fishing out with hooks and line
3. Draining out and drying the pond
4. By poisoning the water

The conventional method of netting does not insure complete removal of all the predatory fishes, as some bottom living species like the murrels, singhi, magur, etc., burrow themselves in mud, and have to be caught by angling. However, complete eradication is possible by dewatering and drying the pond. If this method is not possible or uneconomical, fish poisons have to be used. Poisons are harmful to carp fry as well as to human hence a particular poison has to be used at proper time. Several kinds of chemicals can be used as a fish poisons:

Fish Toxicants

A suitable fish poison is one, which: (*i*) is effective in killing the target organisms at fairly low doses; (*ii*) is not injurious to the people and cattle who may use the water; (*iii*) is easily available; and (*iv*) is, at the same time, economical.

The several kinds of chemicals are used in controlling undesirable fishes. These chemicals may be grouped as: (*i*) plant derivatives; (*ii*) chlorinated hydrocarbons, and (*iii*) organo-phosphates.

(*i*) Poisons of Plant Origin

Derris Powder

Is the most widely used fish poison. It is a plant derivate and contain 5 per cent rotenone ($C_{23}H_{22}O_6$). It is a contact poison and damages the gill membranes, ultimately causing death of the fish. The effect of the poison persists for 4–10 days depending upon the dose used. A dose of 4 parts of derris powder in one million part of water is sufficient to paralyze majority of fishes. Air breathing fishes are killed by 6–10 ppm of derris powder. The required dose is dissolved in a bucket of water, and the solution sprinkled on the surface of the pond, and then splashed about to mix it thoroughly. The poison is very effective in shallow ponds and sunny days and all predatory and unwanted fishes are killed. Poisoning is done about a month before stocking to allow for manuring the pond. After poisoning the fish are seen floating on the surface and can be easily netted out.

Mahau-oil Cake

Mahau-oil cake at 200-250 ppm is used to kill all varieties of predatory and weed fishes are recommended by CIFRI for application in nursery ponds at least a fortnight before stocking nursery ponds with carp spawn. Mahau cake is known to contain about 4 per cent to 6 per cent saponin (morwin). Morwin, soluble in water. According to Nath (1979), the toxic effect of the saponin lasts for 2 to 3 days.

(*ii*) Chlorinated Hydrocarbon

Aldrin ($C_{12}H_8C_{16}$)

It is a product containing not less than 95 per cent by weight of the endo-exo-isomer of 1, 2, 3, 4, 10, 10-hexachloro-1, 4, 4a, 5, 8, 8a-hexahydro 1, 5, 8-dimethano-naphthalene. It is soluble in paraffin and aromatic hydrocarbon solvent. It killed weed fishes and predatory fishes at 0.2 ppm.

Dieldrin ($C_{12}H_8C_{16}O$)

It contains not less than 85 per cent by weight of the endo-exo-isomer of 1, 2, 3, 4, 10, 10-hexa-chloro-6, 7-epoxy-1, 4, 4a, 5, 6, 7, 8, 8a,–octahydrol 1, 4, 5, 8-dimethano-naphthalene. Is soluble in organic solvents. The chemical at 0.01 ppm effectively killed fishes.

Endrin ($C_{12}H_8C_{16}O$)

It is the most poisonous chemical so far tried in India for eradication of fishes. Its formula is 1, 2, 3, 4, 10, 10-hexa-chloro-6, 7-epoxy-1, 4, 4a, 5, 6, 7, 8, 8a-octahydro-exo-5, 8-dimethano-naphthalene. Endrin is soluble in organic solvents. Under laboratory conditions, endrin at a concentration of 0.001 ppm effectively killed all fishes (Chaudhuri, 1960). Its toxic effect lasted for two days. Tafdrin-20, a commercial product of Burmah Shell and Co., containing 20 per cent endrin as its active poison, has been suggested by the CIFRI for use in nursery ponds at a concentration of 0.01 ppm of endrin. The chemical is mixed in water and sprayed over the pond surface. The water is then splashed to mix thoroughly. Fishes die within two hours and float on the surface. Tafdrin should be used 4–6 weeks before stocking. Detoxification of pond water is hastened by using organic manure. For example raw cow dung 18,000 kg/ha quickly neutralizes the effect of endrin.

(*iii*) Organophosphates

In India, three organophosphates, *viz.*, thiometon (Konar, 1969b); DDVP (Srivastava and Konar, 1965 and Konar, 1969a) and phosphamidon (Srivastava and Konar, 1965 and Konar, 1969a) have been found successful for killing fish on experimental basis.

DDVP (0:0-dimethyl = 2, 2-dichlorovinyl phosphate) is commercially available in India as a spray (concentrate 1kg/l) under the name of 'Nuvan 100 EC' (manufactured by Ciba Ltd.). It is soluble in most of the organic solvents. Most of unwanted fishes killed at a concentration ranging from 3 to 30 ppm. Konar (1969a) recommends the use of DDVP for control of trash fishes in nursery ponds. During its use, contact with human skin should be avoided and any accidental splash should be washed off with soap.

Chapter 11
Aquatic Insects and their Control

Nursery ponds designed to rear spawn of major carps are invariably populated with a large number of aquatic insects over greater of the year an especially during and after rains and pray on carp spawn and fry. Most of the aquatic insects, either in their larval and/or adult stages, in addition praying directly upon carp spawn and fry, also compete with later for food.

Eleven orders of class Insecta inhabit the aquatic ecosystem. Among these members of the orders Coleptera, Hemiptera and Odonata are relatively more important. Among Coleoptera members of the families Dytiscidae, Hydrophilidae and Gyrinidae prey upon fish spawn. Dytiscids, known as predaceous diving beetles are carnivorous both in adult and larval stages. They feed on all kinds of aquatic metazoans, the larger species commonly attacking small fish. The larva of *Cybister confuses*, *C. tripunctatus*, *Eretes sticticus*, etc., are observed highly destructive to carp fry. Newly hatched *C. confuses* larva was found to devour 20 advanced fry (24–40 mm in length) within 24 hours (Chaudhuri, 1960). According to Alikunhi *et al.* (1952), grown up larvae of *Cybister* can kill even 10 to13 cm long carp fingerlings. Larva of *Eretes sticticus* are also observed highly predacious on carp fry. The larvae of *Strenolophus rufipes* is observed to kill fish spawn and those of *Dineutus* to feed on small fish (Chaudhuri, 1960).

Among Hemiptera, back swimmers, giant water bugs and water scorpion cause a heavy loss of carp spawn. Back swimmers

(*Anisops bouvieri*) common abound and start attacking and eating carp as soon as the later are released in ponds. The back swimmers are known to attack carp spawn, small fish (Distant, 1906). The food and feeding habits of *Anisops bouvieri*, the most common back swimmer in nursery ponds, have been studied by Gorai and Ray Chaudhuri (1962) who stated that the first two instars of the insect are able to consume only copepods and cladocerans, while the third consume all other plankters. From the fifth instar onwards, a bouvieri feed on fish larvae in addition to, plankton, mosquito and chironomid larvae and may fly nymphs (Julka, 1965). Back swimmers are reported to be capable of killing even 10 to 13 mm long carp fry. For killing, the back swimmers pierce the body of fry by their sharp beak sucking out the body fluids. When starved, they turn cannibalistic.

Water bugs are highly predaceous and feed on frogs and fishes. They secrete a toxic salivary substance which kills the prey in a very short time. Water scorpion, *Laccotrephes griseus*, attacks advanced carp fry and feeds on them. The water stick insects, *Ranatra elongata*, is also a predator of carp spawn and fry. Tripathi and Sharaf (1974) observed *Anisops sp.* (14–15 mm) and *Ranatra sp.* (45-48 mm) to consume 182 and 166 carp spawn in 24 hours.

Among Odonata, dragonfly nymphs are predatory and are observed to feed on carp spawn. Alikunhi *et al.* (1952) reported that a specimen of dragonfly nymph (15 mm long) can swallow 7 carp spawn (6-7 mm in length) within 3 hours in the laboratory. Tripathi and Sharaf (1974) found that the short and long bodied dragonfly nymphs consumed 133 and 125 spawn, respectively in 24 hours.

Control of Predatory Insects

Predatory aquatic insects multiply rapidly between initial poisoning and stocking the pond. Many aquatic insects also fly from pond to pond. Thus, a pond which has been effectively cleared of its insect population, is soon repopulated with insects. For the survival of carp spawn, it is essential that the pond is cleared of its insect population immediately before stocking. Repeated seining of pond with a fine meshed net before stocking does not eradicate the predatory insects.

Effective control of insects is obtained via insecticides. The more common insecticides, however, also affect adversely zooplankton

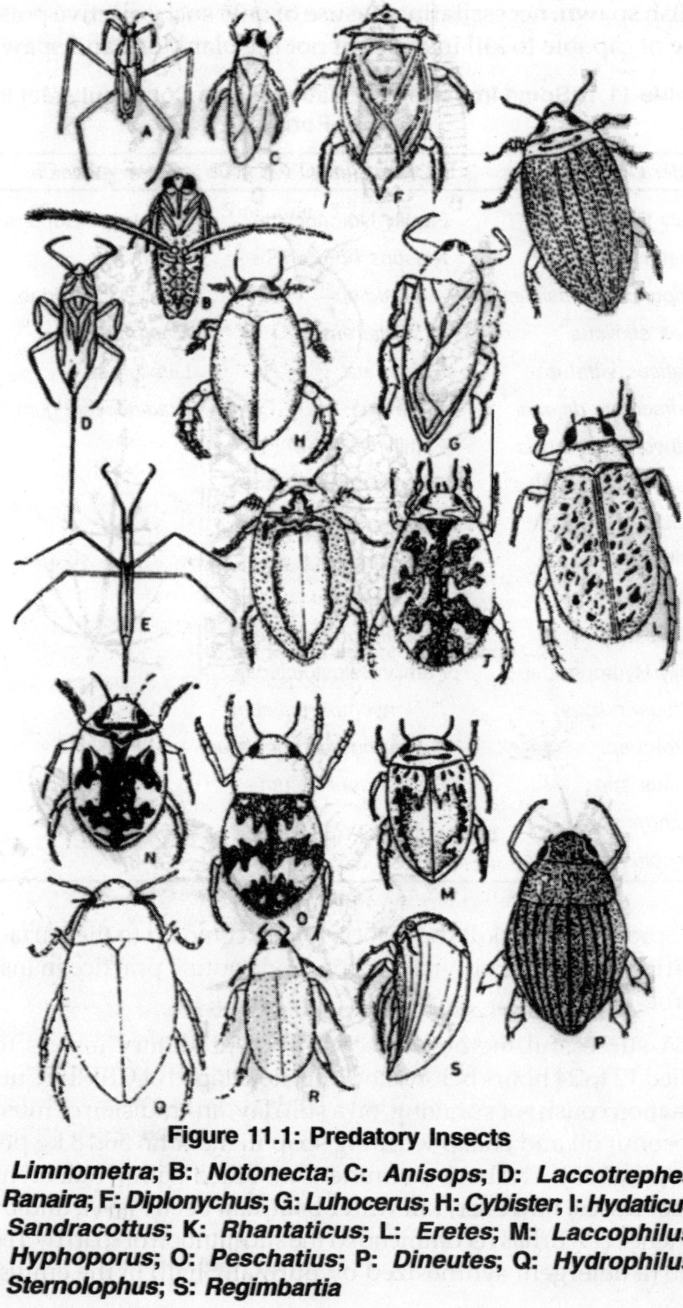

Figure 11.1: Predatory Insects

A: *Limnometra*; B: *Notonecta*; C: *Anisops*; D: *Laccotrephes*; E: *Ranaira*; F: *Diplonychus*; G: *Luhocerus*; H: *Cybister*; I: *Hydaticus*; J: *Sandracottus*; K: *Rhantaticus*; L: *Eretes*; M: *Laccophilus*; N: *Hyphoporus*; O: *Peschatius*; P: *Dineutes*; Q: *Hydrophilus*; R: *Sternolophus*; S: *Regimbartia*

and fish spawn, necessitating the use of only such selective poisons as are of capable to kill insects but not the planktons and spawn.

Table 11.1: Some Important Aquatic Insects Commonly Met in Nursery Ponds

Order Coleoptera	*Order Hemiptera*	*Order Odonata*
Family Dysticidae	Family Notonectidae	S. order Anisoptera
Cybister confuses	*Anisops bouvieri*	*Anax* spp
C. tripunctatus asiaticus	*A. breddini*	Family Libellulidae
Eretes sticticus	*A. waltairensis*	*Macromia* sp.
Hydaticus vittatus	*A. barbata*	*Libellula* sp.
Sandracottus dejeani	*A. sardea*	*Bradinopygia geminata*
Canthydrus laetabilis	Family Nepidae	
Laccophilus parvulus	*Ranatra filiformis*	
Agabus sp.	*R. elongata*	
Family Gyrinidae	*R. digitata*	
Dineutus indicus	*R. varipes*	
Gyrinus spp.	*Laccotrephes griseus*	
Family Hydrophilidae	Family Belostomitidae	
Hydrous indicus	*Diplonychus rusticum*	
Hydrous sp.	*Sphaerodema annulatum*	
Berosus spp	*Belostoma indicum*	
Enochrus spp.		
Laccobius spp.		

Spraying oil to kill the insects which come up to the surface to respiring, is a well known principle and routine practice in insects control.

A successful method for control of predatory insects to be applied 12 to 24 hours before stocking, developed at CIFRI's Cuttack substation consist of spraying, on a still day, an emulsion of mustard or coconut oil and cheap washing soap in the ratio 56:18 kg per ha (Pakrasi, 1953). The back swimmers are killed within half an hour. Higher doses, however, kill water boatman beetle larva and bugs. Chatterjee (1970) has recommended the substitution of soap by Teepol B-300 (a detergent synthesized by Burmah Shall) in the emulsion.

Teepol B–300 is neutral amber–colored liquid, readily soluble in water, the recommended dose of Teepol is 560 ml, emulsified with 56 kg of mustard oil.

Oil of Alexandrian laurel and water dispersible gammexane, commonly known as Hertex U.P. are effective to kill insects. According to Khanna (1957), aquatic insects are killed by gammexane within 1/2 to 11 hours, depending on the concentration of chemical which may vary from 0.62 to 1 ppm. Pure gamma isomer of benzene hexachloride (BHC), soluble in ethyl alcohol, is highly toxic to insects to killing them in about 6 hrs at dose of 0.01 ppm. Carp spawn can tolerate the chemical up to a dose of 0.05 ppm.

A treatment recently developed by the fisheries department, Govt. of Maharashtra, prescribes spraying an emulsion prepared by mixing light speed diesel oil (1 lit.), the emulsifier hyoxid 1011(0.75 mm) and water (40 ml) in the ratio of 1040.75 ml of the emulsion for every 200 m^2 of water surface (Shirgur and Kewalramani, 1967). The use of diesel oil at 50 l/ha mixed with one third to one-fourth of its weight of cheap washing soap has now become a widely accepted practice. The windless day should be chosen for the operation. Insects coming to the surface to breath air come into contact with to oily film which chokes the respiratory tube, killed the insects.

Chapter 12
Aquatic Weeds and their Control

Aquatic weeds are the undesirable plants that grow in water and are more harmful than beneficial for fish culture. Aquatic plants, when in limited quantity are useful and necessary for the ecology of the pond. They form natural food of many species of fish, and fertilize the pond when decayed. They provide shade and shelter to many fish and oxygenate the water. They reduce turbidity and provide spawning beds for fishes.

Aquatic weeds are harmful as they consume nutrients of the pond and obstruct netting operations. They check free movement of the fish and cause oxygen depletion and accumulation of carbon dioxide. Gases like hydrogen sulphide and methane are formed which are harmful to the fish. Algal blooms chock the gills and spoils the water on rotting. Depending on their habits and habitats, weeds have been classified in to the following groups.

(*i*) Floating Weeds

These weeds have their leaves freely floating on the surface of water and roots hanging under the water. They are profusely found in wind protected ponds and are harmful as they shade the pond, *e.g. Eichhorina, Pistia, Lemna, Azolla, Wolffia* etc.

(*ii*) Emergent Weeds

These are surface plants which are rooted in the bottom of the pond, but their leaves float on the water surface or rise above the water level. They prefer shallow parts and shores of the pond, *e.g. Nymphea* (lotus), *Nelumbium, Nymphoides.*

(*iii*) Submerged Weeds

These plants grow under the water surface and may or may not be rooted. The rooted submerged weeds are *Hydrilla, Najas, Vallisneria, Potamogeton* etc. Rootless plants are *Ceratophyllum, Utricularia* etc.

(*iv*) Marginal Weeds

These plants grow along the shore line of the ponds *e.g. Typha, Cyperus, Colocasia, Ipomoea, Sagittaria* etc.

(*v*) Algal Weeds

These are the filamentous algae form a mat on the surface of water. *ex. Spirogyra, Nitella* and *Chara*. These are submerged forms often producing large floating masses commonly referred to as "hair-weeds". Such weeds cause obstruction to free movement of fish. Trapped fish often die unable to come out of the weed.

Aquatic Weed Control

The aquatic weeds can be controlled by three methods. Mechanical, chemical and biological.

Mechanical Control

Mechanical removal of weeds by employing human labour or by machines is the most satisfactory method. Periodical removal of aquatic plants by hand picking uprooting or cutting keeps them in check. The rooted submerged weeds are removed by dragging log weeders fitted with spikes and barbed wire or by cutting then with long handled forks and sickles. Several weeds like the water *Hyacinth, Pistia, Salvina, Azolla* and *Lemna* are the most common floating weeds in ponds and lakes. Complete removal of these weeds by manual labour, drying and burning it up, is considered the most effective method of eradicating it.

Chemical Control

Several chemical herbicides are now available for the control of aquatic weeds. The chief considerations of the use of herbicides in fishery waters are that they should be: (i) effective in killing weeds at reasonably low rates of application, not exceeding 10 ppm for those applied in water and about 10 kg per ha for those sprayed on a

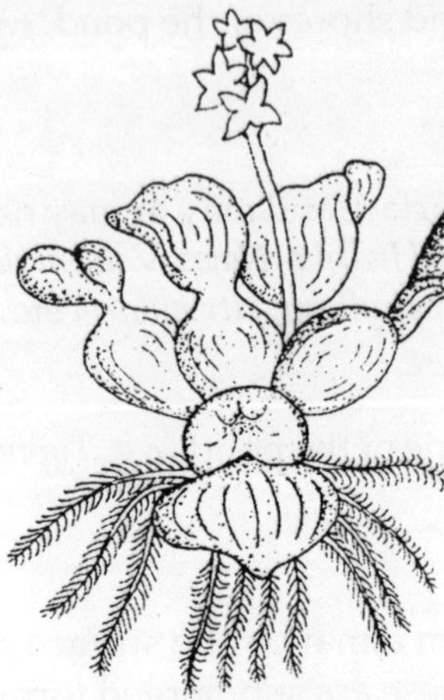

Figure 12.1: ***Eichhornia***

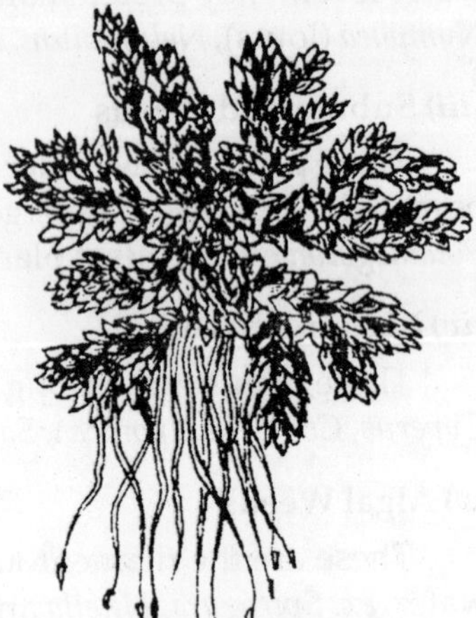

Figure 12.2: ***Azolla***

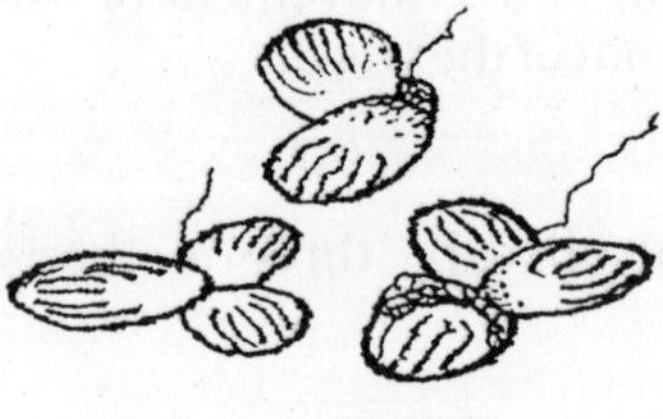

Figure 12.3: ***Lemna***

Figure 12.4: ***Pistia***

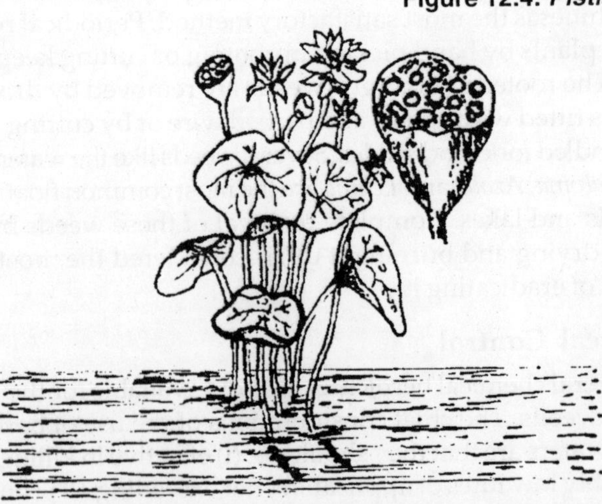

Figure 12.5: ***Nelumbo nucifera***

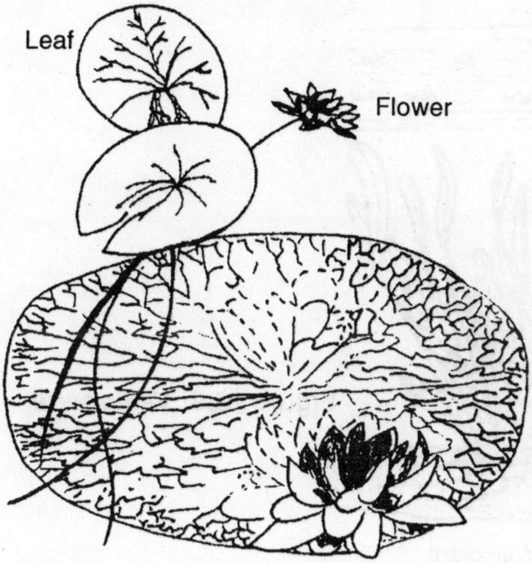

Figure 12.6: *Nymphaea*

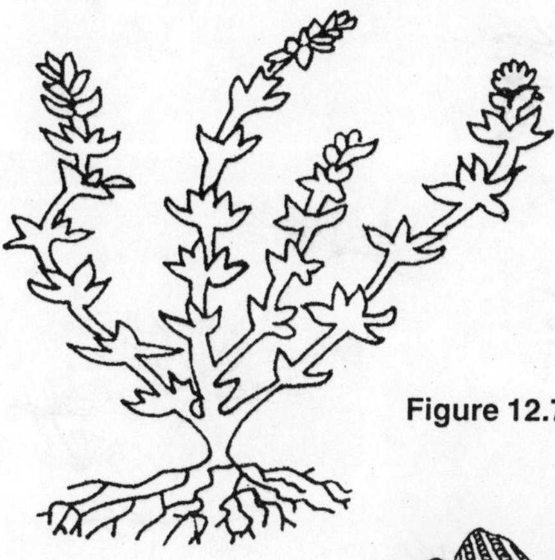

Figure 12.7: *Hydrilla*

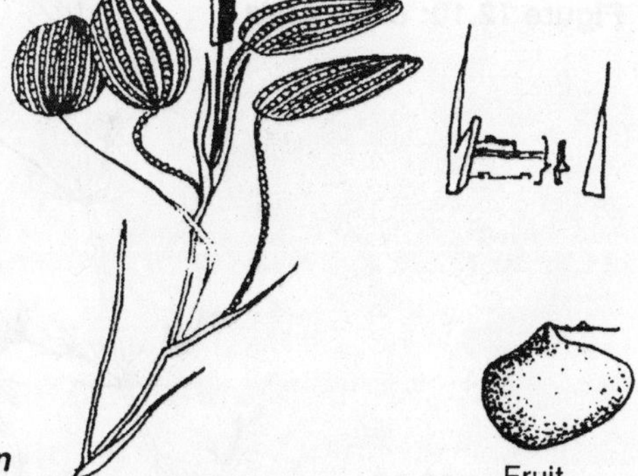

Figure 12.8: *Potamogeton*

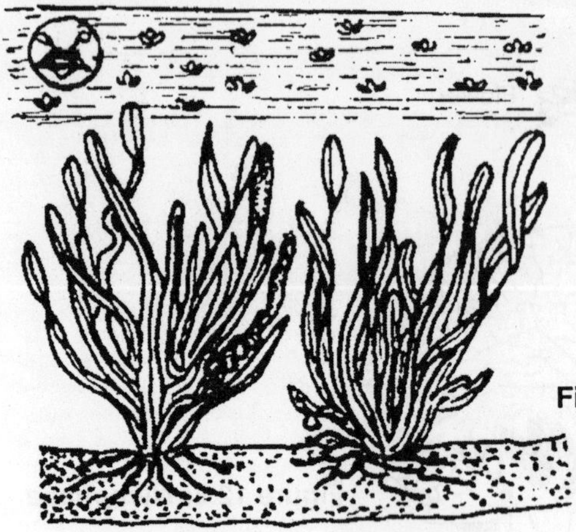

Figure 12.9: *Vallisneria*

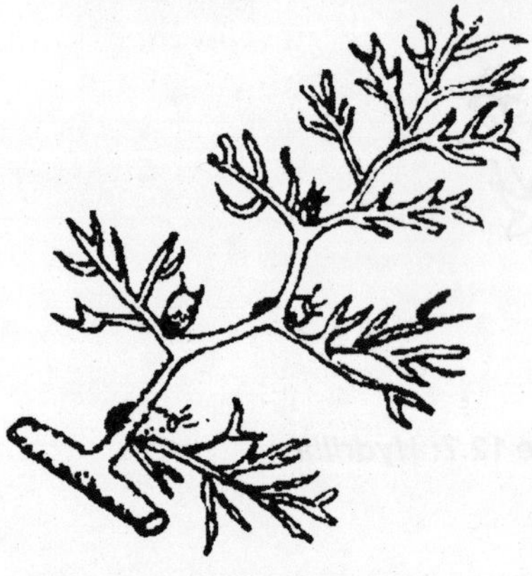

Figure 12.10: *Utricularia*

Figure 12.12: *Ceratophyllum*

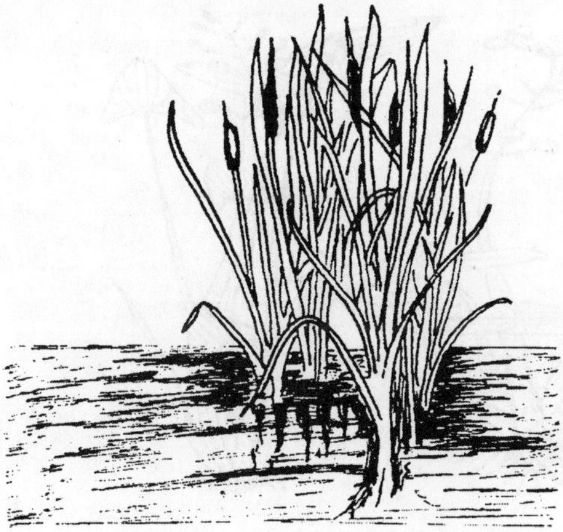

Figure 12.13: *Typha*

Figure 12.14: *Cyperus*

Figure 12.15: *Ipomoea*

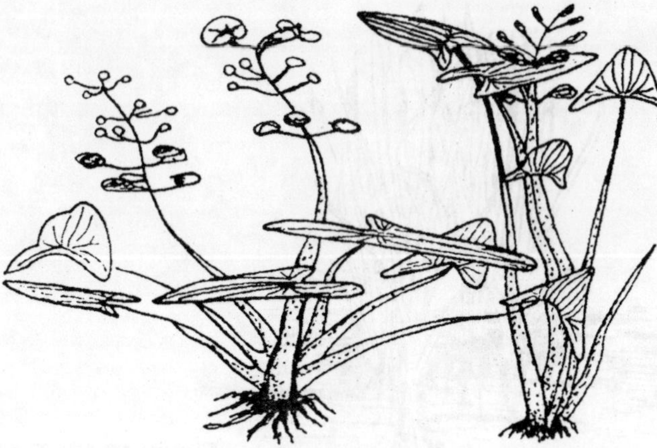

Figure 12.16: ***Sagittaria***

surface area basis; (ii) cheap and easily available; (iii) nontoxic to warm blooded animals so that they could be used freely without any hazard to humans and stock animals; (iv) should not pollute the treated water or leave any residual harmful effects which will adversely affect the fertility of the water and soil; (v) should not involve special costly equipment and constant supervision by technical personnel.

Floating Weeds

It can be controlled by the chemical 2, 4-dichlorophenoxy acetic acid (2,4,-D). This chemical is applied at the rate of 4.5–6.5 kg/ha, and has little harmful effects on fish. Another it kills the plants by deadening the leaves and may also be absorbed by the roots destroying them.

Another herbicide is Taficide-80 (2, 4-D sodium salt 80 per cent) and when used at the rate of 4–6 kg/ha in an aqueous solution of 1–1.5 per cent concentration in combination with 0.25 per cent solution of 'surf' is reported to kill the water *Hyacinth* completely. Simazine (ethyl amino-s-triazine) is also an effective weedicide. 7.5–10 Ibs of chloroxone in 100–200 gallons of water can be sprayed an acre of water. This chemical kills the plants in 2–3 weeks but is harmless to fish and the operator.

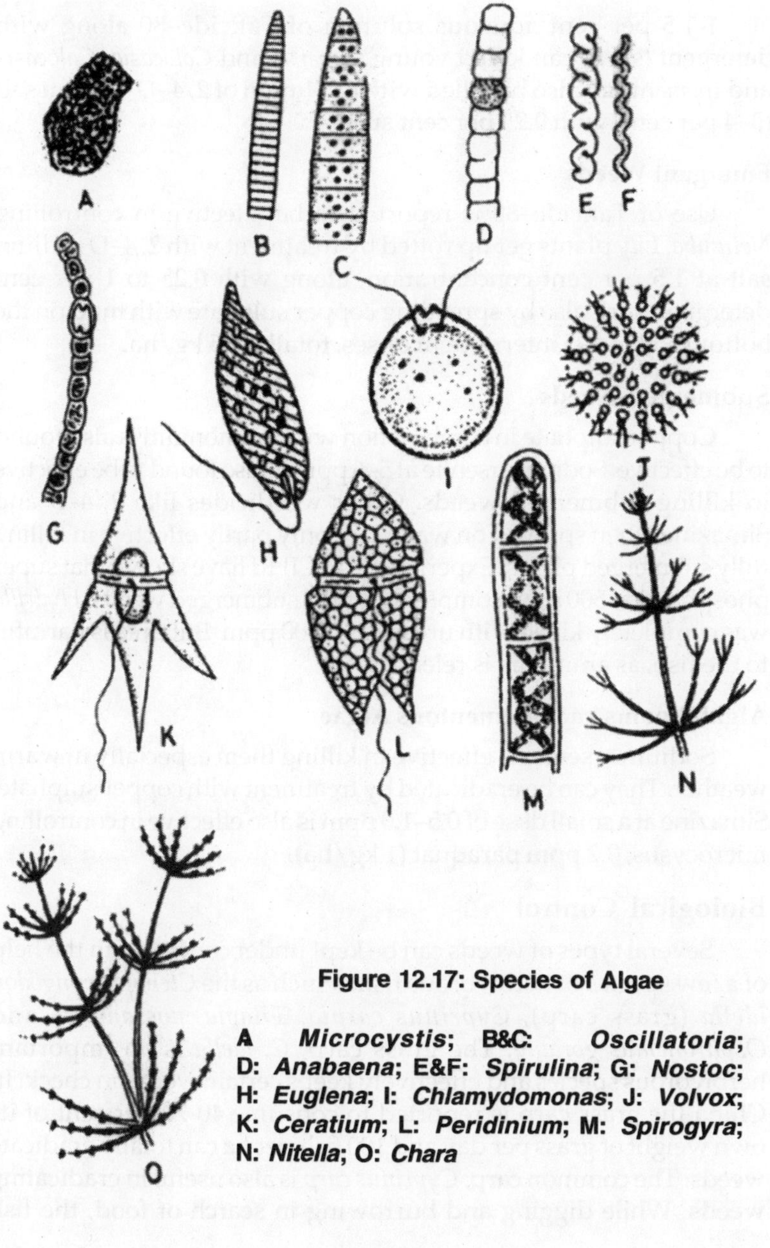

Figure 12.17: Species of Algae

A *Microcystis*; B&C: *Oscillatoria*; D: *Anabaena*; E&F: *Spirulina*; G: *Nostoc*; H: *Euglena*; I: *Chlamydomonas*; J: *Volvox*; K: *Ceratium*; L: *Peridinium*; M: *Spirogyra*; N: *Nitella*; O: *Chara*

Marginal Weeds

1-1.5 per cent aqueous solution of Taficide–80 along with detergent 'surf', can killed young *Cyperus* and *Colocasia*. *Colocasia* and *Ipomoea* can also be killed with a solution of 2, 4–D sodium salt (3–4 per cent) with 0.25 per cent surf.

Emergent Weeds

Use of Taficide–80 is reported to be effective in controlling *Nelumbo*. *Lily* plants get up rotted by treatment with 2, 4–D sodium salt at 1.5 per cent concentration, along with 0.25 to 1 per cent detergent 'surf' also by spreading copper sulphate with mud on the bottom soil in 4–5 intermittent doses, totally 175 kg/ha.

Submerged Weeds

Copper sulphate in combination with ammonium is also found to be effective. Sodium arsenite at 5–6 ppm is also found to be effective in killing submerged weeds. Other weedicides like 2, 4-D and Simazine when sprayed on water, are only partly effective in killing fully submerged plants. Experiments at CIFRI have shown that super phosphates at 500 ppm completely killed submerged weeds. *Hydrilla* was completely killed with urea at 250-300 ppm. But urea is harmful to the fish, as ammonia is released.

Algal Blooms and Filamentous Algae

Sodium arsenite is effective in killing them especially in warm weather. They can be eradicated by treatment with copper sulphate. Simazine at a small dose of 0.5–1.0 ppm is also effective in controlling microcystis, 0.2 ppm paraquat (1 kg/ha).

Biological Control

Several types of weeds can be kept under control with the help of a few species of herbivorous fishes such as the *Ctenopharyngodon idella* (grass carp), *Cyprinus carpio*, *Tilapia mossambica* and *Osphronemus goramy*. The grass carp, *C. idella* is an important herbivorous species and effectively keeps certain weeds in check. In China the grass carp is reported to consume 40-70 per cent of its own weight of grass per day and 100 fish per ha can totally eradicate weeds. The common carp, *Cyprinus carp* is also useful in eradicating weeds. While digging and burrowing in search of food, the fish destroys aquatic weeds by uprooting them. Tilapia and Gourami are also found to be of great help in keeping the submerged weeds under control. Besides fishes, other animals like ducks and geese can also be employed for weed control.

Chapter 13

Fish Diseases and their Control

Like other animals, fish can be subject to many diseases particularly when farming is intensive type. The route cause for the disease in the case of fish is the bad quality of water and its management. When the water is either polluted or condition is deteriorated, the death of all sizes of fish is rapid. But large scale mortality in the case of fish is due to parasitic infections. Health of the fish depends on a good deal on the sanitary conditions prevalent in the water body concerned. The diseases of the fish may be classified as follows:

1. Viral diseases
2. Bacterial diseases
3. Fungal diseases
4. Protozoan diseases
5. Diseases caused by worms
6. Crustacean diseases
7. Nutritional diseases

Viral Diseases

(*i*) Viral Hemorrhagic Septicaemia (VHS)

This viral infection was discovered for the first time by Scaperclaus (1941) and the same was confirmed by Zwillenberg (1965). This particular virus has got RNA genome and measures about 50 nm in diameter. This disease becomes chronic causing very high mortality rate.

Symptoms

Initially there is considerable reduction in the growth rate of fish. The viral infection causes significant changes in the metabolic and biochemical profiles of the tissues. Among the affected organs the damage is maximum in the case of liver and kidney. The important clinical symptoms include abnormal movement and position of fish in water, swollen eyes (exophthalmos), general anaemic condition, reduced hemoglobin percentage, protruded anus, muscular edema, swollen belly, hemorrhages in the air–bladder and in the muscles, presence of neutral or alkaline liquid in the stomach, red intestine, decaying and disintegrating liver, posteriorly swollen kidneys sores on the skin, pale gills, etc.

Treatment

There is no effective treatment of VHS diseases. Fish collection should be from the unpolluted areas. Infected fish should be immediately removed and buried in far away regions. High stocking densities should be avoided. Fat material in the diet should be avoided. But vitaminous food should be preferred. Select the ponds for stocking only after the disinfection process is over. All the ponds selected for stocking should possess a clean bottom.

(*ii*) Infectious Pancreatic Necrosis (IPN)

It is a viral disease causing heavy mortality in the case of young trout. This virus measures 70 nm in diameter.

Symptoms

When fish are infected they revolve on their longitudinal axis. They swim without any sense of direction and finally becomes unable to move. They become dark in color. The eyes protrude and the stomach and the intestine is filled with a whitish fluid. The gall bladder shows signs of necrosis, both liver and spleen are pale.

Treatment

There is no effective treatment for this disease, however, Economon (1963) has suggested the use of povidoneidine which may be useful in controlling the disease.

(*iii*) Spring Viraemia Carp (SVC)

It is caused by Rhabdovirus carpio.

Symptoms

The fish become dark in color. Distended abdomen, hemorrhage in gills and skin and loss of balance are important signs of this infection.

(*iv*) Channel Catfish Virus (CCV)

It is caused by a DNA genome virus which measures 10 nm.

Symptoms

Fish hangs vertically in water showing loss of balance. Hemorrhage of visceral organs may be seen in this disease.

Bacterial Diseases

(*i*) Abdominal Dropsy

The most feared disease of carp is the infectious abdominal dropsy. The bacterium *Aeromonous punctata* is the primary cause of this disease.

Symptoms

During the spring season, the belly of the fish swells by the accumulation of yellow or pink colored liquid in the body cavity. This is the intestinal infection, intestine inflammation, however, liver and kidney also affected. The ulcerative form is recognized by profuse ulceration of skin by the presence of blood areas on the body and the muscle is seriously affected. When the fish are infected they feel pain and unrest which can be their frequent jumping. In some cases fins may be partially destroyed. It has been suggested that viruses are also responsible for causing this disease.

Treatment

Dropsy resistant carps should be selected for culture. Dead fish should be immediately removed. Antibiotic should be mixed either in the feed or they should be injected in a bath, to overcome any possibility of infection. Chloramphenical, Oxytetracycline and Streptophenicol are effective against *Aeromonas punctata*. For mixing the antibiotics with food, a fish with 100 g body weight should be provided 1 mg of it per day. Dipping the fish in copper sulphate solution (1:2000) and also potassium permanganate (5 ppm) for two minute also advisable.

(*ii*) Furunculosis of Salmonids

This is a bacterial disease caused by *Aeromonas salmonicida*. Salmons and trouts are very easily attacked. This bacterial infection is transmitted through the digestive tract and the skin. The bacteria breed in the liver, spleen, kidneys and the blood. *Aeromonas salmonicida* is rod shaped bacterium 2-3 um in length.

Symptoms

Under the skin bloody boils of different sizes can be easily traced. The center of infection is present along the muscular fibers. Inflammation of the intestines, spotted liver, hemorrhages at the bases of the fins, in the gill covers and in the muscles are also important signs of this disease. Mass mortality has been reported in trouts.

Treatment

The disease can be treated with sulphonamides, nitrofurans and antibiotics by mixing them in the food. A daily dose of 10 g sulphamerazine and 3 g of sulphaguanide for a period of 8 days in 100 kg of trout, 10 g of chloramphenicol for 100 g of fry over a period of 10 days and 5 g of it for 100 kg fingerlings or trout over a period of 6 to 10 days. Oxytetracycline may also be administered. Prior to the stocking all the ponds should be disinfected with quicklime or calcium cyanide. Perfect water management in the culture ponds alleviates the bacterial infection.

(*iii*) Fin and Tail Rot

Fin and tail rot affect both adults and young. The infection, during its early stage, appears as white line on the margin of the fin, spreading and imparting frayed appearance to the appendage which eventually putrefies, and disintegrates. The disease is contagious and may cause considerable damage.

Treatment

Bath for 1-2 minutes in 1:2000 solution of copper sulphate till the fish shows distress (Hora and Pillay, 1962); one minute dip in 1:3000 solution of copper sulphate; 1:20000 solution of $CuSO_4$ for 15 minutes treatment of infected portions.

(*iv*) Ulcer Disease

Caused by bacteria has been observed in major carps.

Symptoms

Presence of open sores or ulcers on the body of the fish which increase in size gradually exposing the muscles.

Treatment

The badly infected fish should be destroyed and the pond water disinfected with a 0.5 ppm solution of $KMnO_4$. Dip treatment for one minute in 1:2000 copper sulphate solution for 3–4 days in case of fish showing early stage of infection.

(*v*) Eye Disease

An epidemic eye disease, affecting medium sized and large sized catla, is caused by variant of the bacterium *Aeromonas liquefaciens*. The infected sites are: eyes, optic nerves and brain of the fish.

Symptoms

During the initial stages, cornea of the eye becomes vascularized and later becomes opaque. Subsequently, the eye ball gets putrefied, leading to death.

Fungal Diseases

(*i*) Saprolegniasis

Saprolegnia develops not only in the diseases or dead fish but also develop in the weakened and injured fish also. This parasitic fungus makes a way through the external lesions.

Symptoms

Wolly grey-white or light brown blotches on the fins, eyes, gills or mouth show signs of infection. Tufts of "cotton wool" on the skin (Dermatomycosis) is a common fungal disease in several fish.

Treatment

Potassium permanganate bath 1 g per 100 lit of water for 60 to 90 minutes has been recommended or salt bath 10 g/lit water for 10 minutes or copper sulphate bath 5 g in 10 litre of water are also useful. Fungal development on trout eggs may be stooped by daily formaline bath for 15 minutes 1 to 2 ml of 30 per cent formalin in 1 lit of water. 10 mg concentration of malachite green per litre of water for 15 minutes bath every two days also prove to be highly useful.

(*ii*) Gill rot or Branchiomycosis

Gill rot is caused by the fungal genus *Branchyomyces* in most of the cyprinids and pike. This disease is very common in high stocking densities when there is abundance of phytoplanktons and organic matter. The disease manifests during the summer season.

Symptoms

The gill becomes pale, and on the gills grey regions can be seen. Also the gills show red colored patches.

Treatment

Stocking density should be reduced and also supply of inorganic matter should be avoided. During hot weather even the artificial feeding should be reduced. There is a need for a good supply of freshwater. Application of 200 kg quicklime should be spread per hectare but the *p*H of the water is kept always less than 9. 12 kg of Copper sulphate per hectare for pond with an average depth of 1 m has been recommended. A bath of Benzalkonium chloride 1 to 4 ppm may also be used for 1 hour.

Protozoan Diseases

Numerous protozoans can live on the external or internal parts of the fish bodies and known to cause various diseases.

(*i*) Costiasis

This disease is caused by a flagellate known as *Costia necatrix* when there is prolonged crow dung on tanks. They live on the skin, fins and also on the gills. Costiasis can be easily seen in salmonid fry and young carps.

Symptoms

Slimy secretion appears on the skin, fins and the gills. Seriously affected parts of the skin show red patches. Generally affected gills are brown and partially destroyed. This disease appears when living conditions are favorable such as scarcity of food and the *p*H of water is acidic.

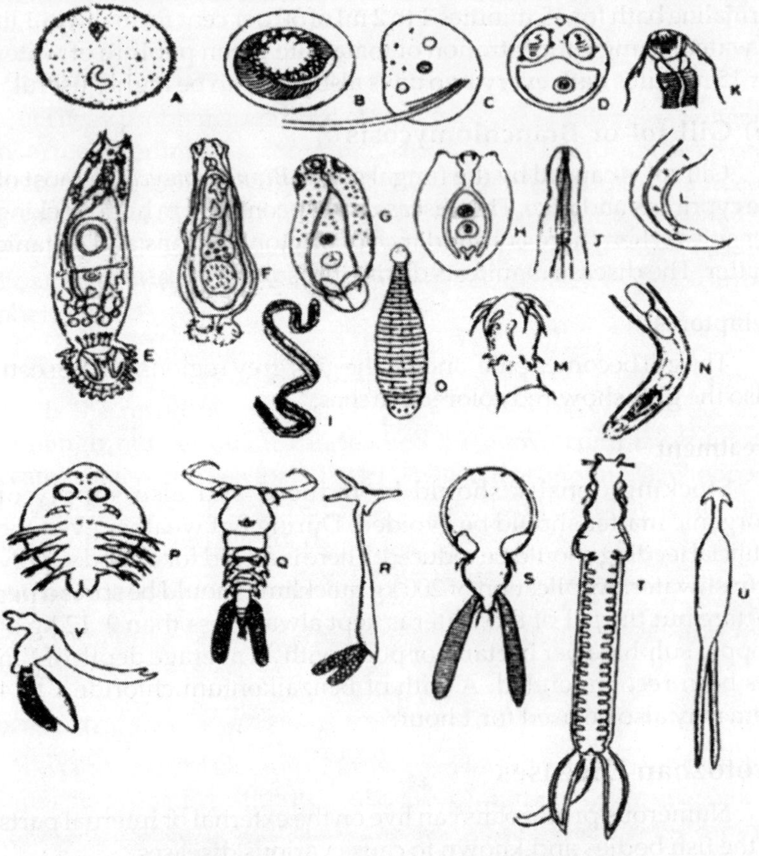

Figure 13.1: Common Parasites of Fish

A: ***Ichthyophthirius*****; B:** ***Trichodina*****; C:** ***Costia*****; D:** ***Myxobolus*****; E:** ***Gyrodactylus*****; F:** ***Dactylogyrus*****; G:** ***Diplostomulum*****; H:** ***Diplostomum*****; I:** ***Ligula*****; J:** ***Philometra*****; K&L:** ***Camallanus*** **(Anterior and posterior regions); M&N:** ***Acanthocephalan, Zelanechinorhynchus*** **(Proboscis and posterior regions); O:** ***Hirudinean, Hemiclepsis*****; P:** ***Argulus*****; Q:** ***Ergasilus*****; R:** ***Lernaea*****; S:** ***Caligus*****; T:** ***Pseudocyonus*****; U:** ***Lernaenicus*****; V:** ***Calvellisa***

Treatment

40 ml of formalin in 100 litres bath for 15 minutes has been advised or 10 g of sodium chloride in litre of water for 20 minutes also proved to be useful.

(*ii*) Whirling Disease

This disease is caused by *Myxosoma cerebralis*. Their spores present on the bottom of the pond are very persistent. The fish are infected by these spores when they are very young. In the intestine of the young fish, the spore liberates sporozoites which, when carried by the blood, settle in the soft cartilages of the head region.

Symptoms

The growth of the fish is much poor. *Myxosoma cerebralis* produces necrosis of the bone and cartilage. Deformation of canals in the ear takes place. This causes loss of balance. The young fish whirl round and fall on the bottom. Some fishes develop black tail after infection in the sympathetic nerve. Deformation of spine, shortening of jaws and gill over, deformation of spinal column and appearance of small cavities in the head are some of the important signs of this disease.

Treatment

Storage tanks should be disinfected with quicklime or calcium chloride or benzalkonium chloride or formalin or sodium hypo chloride.

(*iii*) Ichthyophthiriasis or White Spot Disease

This is caused by a ciliated protozoan *Icthyophthirius multifiliis*. The young parasites attach between the dermis and the epidermis of the fishes. This parasite can cause great losses among carp, trout and other young fishes.

Symptoms

Small grayish spots ranging between 0.1 to 1 mm in diameter are visible on the skin of the fishes. The gills are attacked by these parasites.

Treatment

As parasites are under the epidermis, it is not possible to get rid of the disease. Transfer of fishes from one pond to the another has been suggested. The diseased fish should be carefully removed. Application of 1500 kg quicklime/hectare with an interval of 2 days. Commercial formalin bath (1/4000) or common salt (30 g/litre) bath are also advised.

(*iv*) Myxobolosis (Boil Disease)

The boil disease or ulceration is caused by a sporozoan, *Myxobolus*. Tiny knots are formed in the skin of the common carp, *Cyprinus carpio*.

Diseases Caused by Worms

(*i*) Dactylogyrus

Dactylogyrus is also known as the gill fluke. It is a small flat nematode which can be great danger of young carps. *Dactylogyrus vaslator* is a common species that attack the gills of carps. It is only 1 mm in length. A large number and variety of parasites can attack the young fish at a time.

Symptoms

The gills swell and become grey at the edges and then are partially destroyed.

Treatment

Stocking densities should be reduced. A formalin bath 1 ml/lit of water for 15 minutes for the larger fish and 0.25 to 0.50 ml/lit of water for 30 minutes for the young fish has been recommended. A salt bath of 25 g/lit of water for 10 minutes is also much effective.

(*ii*) Piscicolosis

Fish are attacked by external parasites such as *Piscicola geometra*, the blood sucking leech. The parasite is being provided with a sucker at the end of its body, with the help of which it attaches to the body of the fish and sucks the blood. This is cylindrical, ringed worm not more than 3 cm in its length, 1 mm in diameter.

Symptoms

The parasite makes several incisions in the body of the fish through which fungal attack is feared.

Treatment

The fish should be dipped in the Lysol 1 ml in 5 lit of water for 5 to 10 seconds. Then they are rinsed in the freshwater immediately. Lysol is a mixture of cresol (50 per cent) and soap (50 per cent). Sodium chloride bath 10 g in 1 lit of water for 20 minutes also been recommended.

(*iii*) Ligulosis

Ligulosis is caused by yellowish-white tape worm known as *Ligula intestinalis*. This tape worm is about 15 to 45 cm long and 0.6 to 1.5 cm wide. This disease can be seen frequently among cyprinids found in open water.

Symptoms

This parasite is capable of multiplication and constitute half the total weight of the host fish.

Treatment

The disease can be prevented by eliminating the water foul.

(*iv*) Diplostomosis

This disease is caused by trematode, *Diplostomulum*. "Parasitic Cataract" is formed in the eye of the fish by the lens and as a result, vision is lost.

(*v*) Hirudinea

Leeches have been reported to infect major carps (Southwell, 1915 and Khan, 1944). Saha and Sen (1955) recorded *Hemiclepsis marginata* affecting carps. Leeches feed on the blood of the host fish causing irritation and abnormal movements of the host.

Treatment

Khan (1944) recommended dip in 1:1000 solution of glacial acetic acid and disinfection with 1:10000 potassium permanganate. A single treatment with 5 ppm Gammexane was observed to cure most of the major carps infected with leeches in a sewage irrigated fishery, and a similar second treatment removed all parasites. Gopalakrishnan (1963) suggested dip treatment in 2.5 per cent common salt solution for about 30 minutes.

Crustacean Diseases

(*i*) Argulosis

Argulus (carp lice) is a crustacean parasite of fish. Its color is green-yellow. This parasite attaches to the skin of the fish at the base of the fins by means of hooks and suckers situated under the eyes.

Symptoms

Anemia may result because of wounds, subsequently leading to the death of the fish. *Argulus* discharges toxic substance from the poison gland, which is very harmful for the host fish.

Treatment

The fish should be dipped in the Lysol 1 ml in 5 lit of water for 5 to 10 seconds. Then they are rinsed in the freshwater immediately. Lysol is a mixture of cresol (50 per cent) and soap (50 per cent). A potassium permanganate bath 1 g in 1 lit of water for 40 seconds has also been recommended.

(*ii*) Lernea

This is a highly specialized copepod and is called 'anchor worm' due to its resemblance with worms, having lost all the appendages. It attacks several species of freshwater fishes as the Indian carps, silver carp, gold fish, trout, etc.

Symptoms

The female has hook-like appendages at the anterior end which are used for firm attachment to the host after the parasite has penetrated deep into the tissue of the fish, thus making wounds and holes in the host body. The wounds become infected with bacteria and fungus, causing the death of fish.

Treatment

The infected fish is taken out and the parasite is touched with a brush dipped in a strong solution of potassium permanganate, which kills the parasite. The parasites are killed when exposed to acid water (*p*H 4.0–4.5) within 30 minutes. The pond water can be disinfected with potassium permanganate solution after the fish are removed to another pond.

Nutritional Diseases

Overfeeding may cause several diseases, which may ultimately lead to the death of the fish. Lipoid hepatic degeneration of Salmonids. Enteritis in some fish and hepatoma or rainbow trout are all nutritional diseases. The controlling measures including the avoidance of excess food. The food should be rich in all the essential ingredients which are basically required for the optimum growth of the fish. One day fasting in every week has been recommended.

Chapter 14

Food and Feeding Habits of Freshwater Culturable Fishes

Fishes feed on a wide range of food material and obtain their nourishment from plants as well as animals. The knowledge of food and feeding habits of culturable fish species is essential for intensive fish culture.

Natural food of fishes, Schaperclaus (1933), classified into three groups:

1. "Main food" or the "natural food", which the fish prefers under favorable conditions and on which it survives best.
2. "Occasional food" or the natural food that is well liked and consumed as and when available, and
3. "Emergency food", which is ingested when the preferred food items are not available on which the fish is just able to survive.

Certain microscopic planktonic crustacean groups and rotifers form the "main food" of spawn and fry (15 to 20 mm size range) of the Indian major carps and majority of the other culturable species, with phytoplankton forming the "emergency food". Spawn and fry with a small and short straight intestine appear to digest rotifers and cladocerans fairly rapidly and thrive well on zooplankton. The food and feeding habits of the fingerlings of Indian Major carps differ markedly from those of their hatchlings and fry.

Nikolskii (1963) categorized fish according to the extent of variation in the food consumed by them, such as

1. Euryphagic, feeding on a variety of foods.
2. Stenophagic, feeding on a few selected types of food, and
3. Monophagic, feeding on a single type of food.

The feeding behavior is a species characteristic. Cultured fish often classified according to the trophic niche they occupy in a water-body. Following this system, fish have been grouped into

1. Plankton eating surface feeders. *ex. Catla catla*
2. Column or mid feeders. *ex. Labeo rohita, Labeo bata, Labeo calbasu*
3. Bottom feeders. *ex. Cirrhinus mrigala*

Fishes have become adapted to a wide variety of food. Some of them feed exclusively on plants. Others feed on animals, while a large number of species are omnivorous, deriving their requirements from both animals and plants.

Adult fish, according the character of diet they thrive on, have been classified into herbivorous, carnivorous and omnivorous or but most of them are highly adaptable in their feeding habits, and utilize the readily available food. A few fishes depend for their nutrition entirely on zooplankton and phytoplankton, and are called plankton feeders.

Freshwater fishes can be divided into the following feeding types:

Herbivorous Fishes

Feed mainly on unicellular algae, filamentous algae portion of higher aquatic plants along with some sand or mud. As the plant material in their food is 75 per cent or more of the total gut contents, they are considered herbivorous in habit. Animal food usually varies from 1-10 per cent in their diet. Examples: *Labeo rohita, Labeo bata, Ctenopharyngodon idella, Tilapia mossambica, Osphronemus goramy* etc.

Carnivorous Fishes

Food consists of a very high percentage of animals as crustaceans (Copepods and Daphnids) insects (water bugs, beetles,

Table 14.1: Food and Feeding Habits of Freshwater Culturable Fishes

Species		*Food*	*Feeding Habit*
Catla catla	Fingerlings	– Water fleas, planktonic algae and some vegetable debris	Surface feeder
	Adult	– Crustaceans, algae, plants, rotifers, insects, vegetable debris	Planktophagus
Labeo rohita	Fingerlings	– Vegetable debris, microscopic plants	
	Adult	– Vegetable debris, microscopic plants, decayed higher plants, detritus and mud	Column feeder
Labeo calbasu	Fingerlings	– Vegetable debris, microscopic plants	
	Adult	– Vegetable debris, microscopic plants, detritus and mud	Omnivorous, detritus feeder
Cirrinus mrigala	Fingerlings and	– Decayed plant and animal matter, algae, detritus, planktophage	Bottom feeder
	Adults	– Mud, etc.	
Cyprinus carpio	Fry	– Nauplii	Omnivorous
	Fingerlings	– Rotifers, Cyclops, Euglena, Oscillatoria	Omnivorous
	Adult	– Ostracods, insect including Chironomid larvae, Microcystis, Euglena, Clostredium	Omnivorous
Channa striatus	Fry	– Water fleas, insects, fish fry	
	Fingerlings	– Dipteran larvae, Zooplankton, fish fry	Bottom feeder
	Adult	– Small fishes	Carnivore

Contd...

Table 14.1–Contd...

Species		Food	Feeding Habit
Heteropneustes	Fingerlings	– Insect and worms	Bottom feeder
fossilis	Adults	– Insects, ostracods, debris and algae	Omnivore
Anabas	Fry	– Minute protozoans, animalcules and water fleas	
testudineus	Fingerlings	– Mosquito and other insect larvae, water fleas etc.	
	Adult	– Insects, water-fleas, vegetable debris, fish, etc.	Mainly entomophage
Wallago attu	Fry	– Water fleas, insects, fish fry	
	Fingerlings	– Insects, fish fry and fingerlings	Highly carnivorous &
	Adult	– Small and medium sized fish	Predaceous
Mystus seenhala		– Same as in *Wallago attu*	–"–
Notopterus chital		– Insects, fish fry and fingerlings, small fishes, frogs, tadpole	Carnivorous, predatory
Clarias batrachus		– Insects, worms, crustaceans, fry, larvae, decaying organic matter	Omnivorous
Osphronemus goramy		– Mainly aquatic plants and algae, sometimes insects, larvae	Omnivorous
Tilapia mossambica		– Filamentous algae, aquatic plant materials, insects, crustaceans	Omnivorous

dragon flies, damsel flies), larvae, molluscs, smaller fishes and tadpoles. These species are *Wallago attu, Mystus seenghala, Channa striatus, Notopterus chitala*. Some of these are active predators such as *W. attu, C. striatus, C. marulius, N. chital and M. seenghala*.

Plankton Feeders

Feed on phytoplankton and zooplankton, which they obtain by filtering water through their gill rakers. These are *Catla catla, Cirrhina reba, Hypophthalmichthys molitrix*. They can be both omnivorous and carnivorous.

Omnivorous Fishes

Food of a large number fishes consists of unicellular and filamentous algae, portions of higher aquatic plants, rotifers, insects and their larvae, crustaceans, bryozoans, mud and sand. They are regarded as *omnivorous species* and include *Cyprinus carpio, Cirrhina mrigala, Heteropneustes fossilis, Clarias batrachus, Esomus danricus, Colisa fasciatus*, etc. The food of fishes consists of a varying percentage of plant and animal material, and they form a sort of link between the herbivorous and carnivorous fishes.

Das and Moitra (1955, 1963) have divided freshwater fishes into three groups according to the niche they occupy in the water. These are surface feeders, mid or column feeders and bottom feeders.

The surface feeders are *Catla catla, Hypophthalmichthys molitrix, Puntius ticto, Oxygaster bacaila, Hilsa hilsa*. Some species are neither true surface feeders nor true bottom feeders and mostly depend on the organisms of the mid water. Such species are also called column feeders and includes *Labeo rohita, Tor tor, Wallago attu, Mystus cavassius* etc.

The bottom feeder depend mainly on bottom organisms, fishes included in this group are *Labeo calbasu, Labeo bata, Cirrhina mrigala, Cirrhina reba, Clarias batrachus, Heteropneustes fossilis, Channa* species etc.

Feeding Intensity

The degree of feeding is called feeding intensity and can be ascertained by examining fullness of the stomach and by calculating the gastrosomatic index by the following formula:

$$\text{GaSI} = \frac{\text{Weight of the stomach content}}{\text{Weight of the fish}} \times 100$$

The feeding intensity of the fish varies with the season and is related with maturity, spawning and the availability of food items.

Chapter 15
Fish Nutrition

In intensive fish culture practices the objectives is to maintain optimum density of fish per unit area of water by adopting techniques relating to polyculture, multiple stocking, stock manipulation etc. Under such conditions of fish culture, the natural protein component of food organisms present in the environment will not meet the needs of the growing fish biomass, there by necessitating supplementation with protein rich needs. The development of artificial feeds mainly depend on the studies relating to basic nutrition and physiology. For economic reasons such investigations should aim to find out the minimum level of protein which satisfies the amino acids requirements of the species for optimum inherent capacity for growth, adequate supplementation by carbohydrates to serve as dietary calories and luxurious supply of vitamins and minerals for necessary stimulation of protein digestion.

Nutritional Requirements

Fast growing varieties of fish are generally chosen for culture practices. The practice is to provide some starchy foods to several fast growing fish to serve as dietary calories. To meet the dietary demand one should know the nutritional requirements of fish such as proteins, carbohydrates, fats, vitamins, micronutrients, etc., besides the knowledge relating to digestibility and utilization of the compounded feeds by the fish for yielding protein as the final metabolized product in the intensive fish culture practices.

Carbohydrates

Carbohydrates are the most abundant and relatively less expensive source of energy in animal aquaculture. These may consist of easily digested sugars to complex cellulose which is difficult to digest. Most of the cultivable fishes like carps and mullets are omnivorous taking in considerable amount of vegetable mayyet and are therefore, well adopted physiologically to digest starch. Digestibility of starch is reported to be 30-90 per cent. Rice bran and wheat bran which are the main starchy diets used for cultivable fishes are found to be highly digestible. The ratio of protein to carbohydrate in the feeding of 1:7 or 1:8 which gives a wide scope to utilize feeding of cheap carbohydrate diets as long as for protein in the natural food is sufficient for growth. While formulating the balance diets, carbohydrate and protein ratio needs a careful manipulation so as to spare the proteins for growth and carbohydrate to serve supplying the dietary calories.

Proteins and Amino Acids

Proteins are essential components of both the cell nucleus and cell protoplasm and accordingly account for the bulk of the muscle tissues, internal organs, brain, nerves and skin. Fish were fed a balance diet containing graded levels of high quality protein (casein: gelatin mixture supplemented with crystalline amino acids to stimulate the amino acid profile of whole egg protein).

Dietary protein is the main source of nitrogen and essential amino acids. It is also the most expensive source of energy in artificial diets. In nature, carnivorous fish consume food which are about 50 per cent protein.

Essential Amino Acids

Fish requires 10 essential amino acids arginine, histidine, isoleucine, leucine, lysine, methionine, phenylalanine, threonine, tryptophan and valine. Qualitative requirements for amino acids are determined in fish by feeding a purified diet composed of crystalline amino acids as a control diet and feeding test diets. Test diets that produce no growth or markedly less than the control represent amino acids that are essential to the fish. Quantitative requirements for essential amino acids are determined by feeding graded levels of one amino acid at a time in a test diet is usually similar to that fish eggs or of the fish muscles.

Dietary imbalances of amino acids can cause reduced performance by animals through amino antagonism or toxicity. When some amino acids are fed in excess of their required levels, they cause an increase in the requirement for other structurally similar amino acids, or amino acid antagonism. In some instances, however, dietary excess of certain amino acids are directly toxic and their negative effects cannot be ameliorated by addition of other amino acids; this in amino acid toxicity. Fish diet containing practical feed stuffs, such as grain by products, oil meals and animal by products are not likely to be so seriously imbalanced, but special diets could be.

The requirements are the highest in the initial feeding of the fry and decrease as fish size increases. For maximum growth, young fish require between 40 to 60 per cent of their diet as proteins.

Table 15.1: The Gross Protein Requirements of a Number of Fin Fish

Species	*Crude Protein Level in Diet for Optimum Growth (g/kg)*
Rainbow trout	400-600
Common carp	380
Salmon	400
Eel	445
Grass carp	410-340

Protein requirements are influenced by water, temperature, body size, stocking density, oxygen levels and the presence of toxins. Proteins has more net energy for fish. Fish are among the most efficient of all animals in converting feed energy into high quality protein. Phytoplankton and zooplankton contain high percentage of protein (40-60 per cent) and there is reason to believe that the protein requirements of plankton feeding species are also similarly high.

Amino Acids

Amino acids are described as the building blocks of proteins and about 23 of them have been isolated from natural proteins. 10 of these amino acids are considered indispensable for fish, these are arginine, histidine, isoleucine, leucine, lysine, methionine, phenylalanine, threonine, tryptophan and valine, alanine, aspartic acid, cystine, glutamic acid, glycine, proline, serine and tyrosine are considered to be the non-essential amino acids. Casein supplemented with 6 AA's gave feed conversion ratios similar to isolated fish proteins as a dietary source. It was demonstrated that soybean meal supplemented with five or more AA's, increasing methionine and lysine.

Table 15.2: Some Quantitative Essential Amino Acid Requirements of Certain Species of Fish in g/kg Diet

Amino Acids	*Salmon*	*Eel*	*Carp*
Arginine	24	17	16
Histidine	7	8	8
Isoleucine	9	15	9
Leucine	16	20	13
Lysine	20	20	22
Methionine	16	12	12
Phenylalanine	21	22	25
Threonine	9	15	15
Tryptophan	2	4	3
Valine	13	15	14

Lipids and Essential Fatty Acids

Lipids are a group of fat soluble compounds occurring in the tissues and broadly consists of fats, phospholipids, sphingomyelins, waxes, esters of glycerol and are the principal form of energy storage. Fish in general contain more w_3 than w_6 PUFA (polyunsaturated fatty acids), but freshwater fish appear to have higher levels of w_6 fatty acids than marine species. The lipid contents of most commercial aquaculture diets are less than 10 per cent mainly due to processing problems. Dietary lipids may be used to spare the more valuable protein for growth. In particular, free fatty acids derived from triglycerides are the major aerobic fuel source for energy metabolism of fish muscle.

Lipids serve as biological carriers for the absorption of the fat soluble vitamins A, D, E, and K. From feed technology view point, lipids act as a lubricants for the passage of feed through pellet diets, as substance which reduce the dustiness of feeds and play a role in feed palatability.

Essential Fatty Acid Requirements

In view of inability of animals to synthesize *de novo* fatty acids of the n-6 and n-3 series, these fatty acids must be supplied in a ready made form within the diet. It is generally believed that the dietary requirement of fish for n-3 series EFA, over n-6 series is fundamentally due to the low water temperature of their aquatic environment. In fact, the lower the water temperature, the greater the incorporation of n-3 series PUFA in the tissues. In general, cold water freshwater fish have an exclusive requirement for n-3 series PUFA (18:3 n-3, 20:5 n-3, 22:6 n-3) in their diet.

Table 15.3: Dietary Essential Fatty Acid (EFA) Requirement of Fish

Fish	*Requirement*						
Rainbow trout	1%	18:3	n-3				
or	1%	HUFA	n-3				
Ayce	1%	18:3	n-3				
or	1%	20:5	n-3				
Common carp	1%	18:3	n-3	+	1%	18:3	n-3
or	0.5	—	1%	HUFA	n-3		
Eel	0.5%	18:2	n-6	+	0.5%	18:3	n-3

Vitamins

Vitamins are a heterogenous group of organic compounds essential for the growth and maintenance of animal life. Approximately 15 vitamins have been isolated from biological materials, their essentiality depending on the animal species, the growth rate of animal, feed composition, and the bacterial synthesizing capacity of the gastrointestinal tract of the animal.

Water Soluble Vitamins

The water soluble vitamins include eight well-recognized members of the vitamin B complex, are required in small amounts in the diet but play important roles in growth, physiology and metabolism.

Thiamin

Thiamin pyrophosphate acts as a coenzyme for several

metabolic decarboxylation and transketolation reactions. Thiamin deficiency affects the central nervous system, hypersensitivity to disturbance, loss of equilibrium, and convulsions. Channel catfish fed diets containing 40 per cent non heated fish viscera developed thiamin deficiency in 10 weeks, but when the fish were fed an additional diet containing thiamin in a desperate meal daily, no deficiency occurred.

Riboflavin

A number of riboflavin deficiency signs have been observed photophobia and cataracts have been found in several fish deprived of riboflavin. Necrosis of head kidney was observed in carp. Bran or polishing from grains and distillery byproducts are fair to good sources.

Niacin

Most fish show niacin deficiency signs rather quickly. Skin lesions are a common deficiency sign in fish. Niacin deficiency can be demonstrated rapidly in fish fed tryptophan in the diet. Because of the presumed limited availability of niacin in grains and oil seed meals to fish, bits supplementation of practical fish feeds is recommended.

Pantothenic Acid

Pantothenic acid is a dietary component essential for fish. A dietary deficiency results in growth failure. Gill lamellae, kidney tubules, and acinar cells in pancreas, all of which carry on a high level of metabolic activity, are all sensitive to pantothenic acid deficiency. Pantothenic acid is widely found in commercial feed stuffs, but the level and availability in processed feed are likely to be lower than the requirement for most fish. Supplemental pantothenic acid is recommended for commercial fish seeds, especially for small fish.

B_6 (Pyridoxine)

Vitamin B_6 is a dietary requirement of all fish. It is widely found in feed stuffs of plant and animal origin. Because fish are fed much higher protein diets than land animals, it should be expected that the B6 requirement would be higher for fish. Vitamin B_6 is involved in the metabolism of carbohydrates and lipids. It is essential for the synthesis of heme (in hemoglobin), and in the synthesis of serotonin from tryptophan, which may explain why B_6 deficiency causes nervous disorders.

Deficiency signs in fish develop quickly young carps in 4 to 8 weeks. These include nervous disorders, such as hypersensitivity to disturbances: poor swimming co-ordination: convulsions; and tetany when handled. Common carp show edema, exopthalmos and skin lesions. The requirement for fish is 3 mg/kg to 10 mg/kg of diet.

Biotin

Biotin functions as a coenzyme in carboxylation reactions. The dietary requirement for fish is low, about 0.25 mg/kg of diet.

Folic Acid

Folic acid deficiency could not be demonstrated in common carp. Poor growth was seen in folic acid deficient eel and channel catfish, combined deficiencies of folic acid and vitamin B_{12} enhance anemia in fish. Grains, oilseed meals, and animal byproducts are all good sources. One to five milligrams per kilogram of diet is sufficient for young salmonids.

Vitamin B_{12} (Cyanocobalamin)

Vitamin B_{12} functions as a coenzyme in a number of metabolic reactions. Vitamin B_{12} deficiency causes microcytic anemia and fragmented erythrocytes with low haemoglobin values in salmonids. Anemia was not found in vitamin B_{12} deficient channel catfish or carp, but slight reduction in growth occurred in the catfish. Internal synthesis is a sufficient source.

Vitamin C (Ascorbic Acid)

Deficiency of vitamin C showed deformed spinal column, distorted gill support cartilage, hemorrhagic areas under the skin, depigmentation, and slow wound healing in several fish species. Most animals can synthesize vitamin C in sufficient quantity for normal growth and function, but many fishes cannot because they lack the enzyme L-glucooxidase for synthesis of vitamin C from glucose.

Dietary requirement for vitamin C by fish seems to decrease with age. Fish fed diets deficient in vitamin C have reduced resistance bacterial diseases. 30 mg/kg of vitamin was sufficient to prevent scurvy in the fish and 150 mg/kg significantly reduced mortality.

Inositol

In several fish species reduced growth, anemia, fin erosion, and slow rate of gut emptying have been reported with inositol deficiency.

Fat Soluble Vitamins

Vitamin A

Vitamin A is its role in vision. Common carp showed deficiency signs of light skin color, fin and skin hemorrhages, exophthalmos, and deformed gill opercula. Thus supplemental vitamin A may be added to fish feeds.

Vitamin D

Vitamin D is essential for maintaining homeostasis of calcium and inorganic phosphate.

Vitamin E

A major function of vitamin E as a metabolic antioxidant, with a specific role in preventing oxidation of unsaturated phospholipids in cellular membranes, such as erythrocytes, and subcellular membranes such as mitochondria.

Vitamin E deficiency signs in several fish species include fatty livers and ceroid (dark lipoid) bodies in liver. "Sekoke disease" in common carp characterized by thinning of flesh on the back, but corrected by supplementing the diet with vitamin E.

Vitamin K

Vitamin K is necessary for normal blood clotting in all animals including fish. Fish meal and alfalfa meal of a good sources of vitamin K. Menadione sodium bisulfite or menadione dimethyl pyrimidinal bisulfite are synthetic sources used in commercial feeds.

Minerals

The mineral requirements of fish have been studied only sparsely. Fish absorbs dissolved minerals from the water. Mineral metabolism in fish is osmoregulation or maintenance of osmotic balance between body fluids in the fish and the water around the fish. Other biochemical functions of minerals in fish are similar to those in warm blooded animals. Some minerals are constituents of hard tissues such as bone, fins and scales and some components of soft tissues, such as sulfur in protein and iron in hemoglobin.

Berg (1968) provided data that indicated that goldfish obtained 50 per cent to 80 per cent of their calcium from the water when fed a calcium-adequate diet. Fish required a dietary source of phosphorus to meet their relatively high metabolic requirement because levels dissolved phosphorus in natural waters are relatively low.

Magnesium

About 70 per cent of the magnesium in a fish's body is in the hard tissues other functions of magnesium are as an enzyme activator in carbohydrate metabolism and in protein synthesis. Deficiency causes flaccid muscle in fish. Fish in freshwater, which only contains 1 mg/L to 3 mg/L of magnesium, required 0.025 per cent to 0.07 per cent magnesium in the diet.

Iron

The principal role of iron in a fish body is as a component of hemoglobin. Another role is as a component of the cytochrome enzyme system that regenerates ATP in cellular oxidation. A deficiency of iron can cause anemia. Dietary deficiency of iron causes microcytic anemia in fish. Dietary requirement of iron by channel cat fish is 30 mg/kg of diet.

Copper

Copper is involved with iron absorption and metabolism. When the diet is deficient in copper, iron levels in body tissues decrease. Copper functions in hematopoiesis (hemoglobin formation) and in several enzyme systems, such as cytochrome C oxidase and tyrosinase. It is essential in bone development. Common carp require approximately 3 mg of copper per kg of diet for normal growth.

Iodine

Iodine is a component of thyroxin, the thyroid hormone that regulates rate of metabolism. Deficiency of iodine results in hyperplasia of the thyroid gland, or goitre, in fish. The minimum requirement of fish for dietary iodine has not been defined; however, 1 mg/kg to 5mg/kg feed has been found to be an adequate level. Fish meal is a rich source of iodine fish feeds.

Zinc

Zinc has a several functions. It serves as a cofactor in several enzyme systems, including carbonic anhydrase found in red blood cells, enzymes in protein digestion, and enzymes in carbohydrate catabolism. It plays a role in preventing teratinization of epithelial tissues. Insulin is stored in zinc complex. Zinc deficiency in common carp caused slow growth; loss of appetite high mortality, and erosion of the skin and fins. Dietary levels of 5 mg/kg allowed maximum growth rate, but 15 mg/kg to 30 mg/kg were required to prevent all deficiency signs.

Manganese

Manganese functions as cofactor in several enzyme systems, including those involved in synthesis of urea from ammonia, amino acid metabolism, fatty acid metabolism and glucose oxidation. Manganese deficient diets caused depressed growth in common carp and rainbow trout, and abnormal tail growth and shortening of the body in the latter. Supplementation of the diet to bring the level of manganese to 12 mg/kg to 13 mg/kg improved growth in both species and prevented abnormalities in rainbow trout.

On an average, the requirements of carps (per cent of diet) is as follows:

Table 15.4: Nutritional Requirement of Carp

Stage	*Protein*	*Fat*	*Carbohydrate*
Adult	40 per cent	20 per cent	10 per cent
Fingerling	45 per cent	6 per cent	26 per cent

Artificial Feed; Pellets and Feeders

In intensive (high density) fish culture use of artificial feed is very important and essential. The artificial feed must meet the total food requirements of the farm fish for maximum growth and metabolism and therefore has to be suitably formulated as balanced, energy rich diet that serves fully for supplementary feeding. Selection of correct ingredients of the feed is essential which in turn will depend upon the quality and quantity of available natural food, feeding habit and growth requirement of the farm fish as well as cost effectiveness and availability of ingredients. The composition of ingredients varies according to the species, age and weight of the stocked fish.

The artificial feed formulation possible effect on the proximate composition (protein and fat) of the fish flesh. Nutritive value of the feed is not only determined by the contents but by the digestibility and assimilation ability of the nutrient contents. Digestibility of nutrients varies from fish to fish. The composition of the feed must match the digestibility coefficient of the farm fish. Digestibility of the nutrients and energy are to most important criteria in evaluation of the feed. High carbohydrate content may adversely affect protein digestion.

From the stand point of growth, the feed must ensure large amount of fat, moderate amount of protein, large amount of water and traces of vitamins and minerals. Fixed feeding will promote better growth if fish are fed on protein-rich diet. High energy diet and high protein diet must balance each other to give correct energy-to-protein ratio. Fish being a cold-blooded need less energy for locomotion and equilibrium in water, but eat to satisfy their metabolic energy requirement. Thus, the artificial feed must meet the energy budget which is balanced between energy input and energy expenditure. In intensive fish farming, if there is a deficiency of diet it will result in ailment and poor growth of the farm fish and if there is an excess it will be an economic waste and may cause pollution, in addition.

Ingredients for Carp Feed

The common ingredients used in artificial feed for Indian carps (indigenous and exotic) include:

1. Powdered algae (*Spirulina*)
2. Aquatic weeds
3. Grasses, legumes, foliage and seeds
4. Oil cakes (ground-nut, coconut, mustard, sunflower)
5. Bran (rice and wheat)
6. Wheat flour, broken rice and roasted gram powder
7. Cattle blood, meat meal, insect pupae
8. Poultry and milk by-products
9. Trash fish, fish silage and fish meal

Selection of ingredients for the feed is based more on availability than nutritional value, to make the feed cheap. Combination of ingredients in carp-farming are as follows:

1. Oil cake and rice bran in ratio of 1:1 is good for carp spawn and fry to supplement plankton and other micro fauna.
2. Protein plus carbohydrate in ratio of 50:30 is good for fingerling and stocked carp.

Composition of Carp Feed

The major components of the dry matter, apart from moisture, are (1) ash (in organic content) and (ii) the organic matter (proximate components, vitamins, minerals and crude fibers). Presence of adequate amount of fibers help in improving pellet quality. Standard feed formulation has 40 per cent of protein, 10 per cent of fat and 1 per cent of vitamin, apart from mineral mixtures, water, additives etc.

Types of Diet

There is a great variety of fish diet dry, moist or wet, floating or non floating. It may be in the form of granules, crumbles, balls, cakes, flakes, pellets or even mash and paste.

Pellets

Pellet is a dry feed. As compared to wet feed, it is easy for mechanized (non-manual) distribution and its storage time is also longer (3 months). Pelleting is done of the grinded and mixed ingredients by means of exposure to moist heat. Compressed pellets are produced by exposure to dry steam followed by moist heat (80°C plus 16 per cent humidity). To make them of lower density so that they may float or sink slowly in water, a still higher temperature of 125–150°C and higher humidity levels (20-25 per cent) are required. Presence of fiber and carbohydrates may be necessary for pellatization. Top dressing of pellets is done by final spraying of fat or oil on the pellet coat.

Various designs of mechanical feeders are used for distribution of pellet into the pond which ensures intermittent falling of the pellets. A timer may be attached to the device for regulation of the intervals between feeding. Falling of pellets, their amount and frequency is, however, mechanically controlled in a variety of way.

Feeding Schedule

Table 15.5: Feeding Schedule of Spawn for 15 Days of Rearing

Period	Feed Per Day in Terms of Weight (g) of Spawn at the Time of Stocking	
	Alikunhi (1957)	Hora and Pillay (1962)
First 5 days after stocking	Double	Equal
6th to 10th days after stocking	Three times	Double
11th to 15th days after stocking	Four times	Three times

Feeding Schedule of Carp Fry to Fingerlings (Lakshmanan *et al.,* 1968)

Mustard oil cake and rice bran in 1:1 ratio by weight at 1400 kg/ha. During first month, equivalent to initial total weight of fry stocked daily; during second and third months, twice the initial weight of fry stocked.

For Indian major carps, a mixture of rice bran and mustard oil cake or ground nut oil cake (1:1 by weight) is considered to be the best. Artificial feed is given daily @ 2-3 per cent of the total fish biomass. It is either sprayed over the water surface or kept as thick paste in shallow, wide mouth earthen pots or trays at a few spots in the pond. Feeding is usually done in the morning or evening, and should not be changed suddenly.

Chapter 16

Productivity of Freshwater Fish Pond

The productivity of a water body connotes a characteristic where by living substance is manufactured through interaction of the constituents of the natural environment. Some ecosystems like ponds and lakes are essential self-sustaining. The fundamental step in the operation of a self-contained ecosystem are:

1. The reception of energy
2. Production of organic materials by producers
3. Consumption of this material by consumers and its further elaboration
4. Decomposition to inorganic compounds, and
5. Its transformation in to suitable forms for the nutrition of the producers.

In a self-sufficient ecosystem, all these step are contained within it. The nonliving constituents are light energy and inorganic nutrients, both required for the growth of photosynthetic plants.

Concept of Productivity

Three measures of productivity, namely:

1. Standing crop
2. Rate of production, are in common use and from the core of the concept of productivity in an ecosystem.

(*i*) Standing Crop

The standing crop is the total biomass of the organisms existing in the area at the time of observation and may be expressed as number of individual, as biomass, as energy content, etc. the term standing crop has been redesignated as the usable stock by Ricker (1958) in case of fish populations. The standing crop, as a measure of productivity. The measurement of standing crop can only reveal the population density of individuals in the ecosystem at a given time. Thus it becomes essential to know the rate at which observed density of the community is produced. The relation of the size of standing crop to the rate of growth of population there by follows. The abundance of a population depends up on many variables operating in an ecosystem. A population is defined by as a group of organisms occupying a given space at a particular time and may consists of a single or more species. All natural populations possess an innate capacity for increase which, at its maximum rate, is called the biotic potential. Biotic potential implies those characteristics and abilities inherent within an organisms which enable it to survive and reproduce, and is the sum total of its capacities which determine its relative success in solving all its problems of maintenance. An absolute biotic potential of any species is probably rare in nature due to limitations imposed by environmental factors, the environmental resistance. Therefore, the growth and success of the population of a species depends up on the degree of equilibrium between biotic potential an environmental resistance.

The actual, or realized rate of growth, is determined by the initial number or individual plus additions (natality) minus the number lost (mortality) per unit of time. In other words, the rate of population growth is geometric and is represented by the relationship:

$$dN/dt = rN$$

where, N= number of individuals; t = time; r = rate of natural increase of population (*i.e.* difference between natality and mortality at a given moment); dN/dt gives the average rate of change of N unit time.

The most common form of growth pattern of a population, under natural conditions, is the logistics curve termed as asymptotic growth. The maximum standing crop which can maintain itself indefinitely in an area is defined as the carrying capacity of that area. Within a given population the relative number of individuals of each group is important in determining the growth from. The importance of age distribution of population lies in the sum total of the natality–mortality values for each of the age groups in the population. The stability of the population structure depends upon nearly equal rate of natality and mortality in each of these age groups, deviation from which result in increase or decrease of population size.

Rate of Removal

Various aspect of the biota removed or harvested from an ecosystem from the second measure of productivity. It refer to the yield or harvest from an ecosystem from the second measure of productivity. It refers to the yield or harvest from an ecosystem per unit of time. Its importance consists in as a criterion to decide whether the substance can be removed (harvested) from an ecosystem without interfering with the natural biogeochemical cycles.

The emigrations of animal of plant life, predation by terrestrial animals, outflow of organisms through effluent streams, loss of nutrients of sediments, commercial harvesting of fish, etc., are some of the ways of removal. Fishing mortality in respect of fish population is only a portion of the total amount of material removed from the area. However, if a community is to maintain itself, the loss of material must be compensated by adequate replenishment. A suitable balance has to be maintained among all members of the ecosystem.

The natural supply of materials entering the ecosystem may be augmented artificially by adding fertilizers or by stocking. In addition, the degree of crowding of the desired species may be adjusted to the optimal level, and the competitors of this species may be removed. The population of the species to be harvested should also be adjusted to the level of abundance at which its growth rate will be as high as possible without disturbing the equilibrium of community. An aspect of yield of great importance is catch or yield per unit effort, which is defined as the catch of fish in numbers or in weight taken by a definite unit of fishing effort (Ricker, 1958.)

Rate of Production

This aspect of productivity relates to the amount of organic substance synthesized in a certain space per unit of time. It involves the cycle of growth, reproduction, death and decomposition of organisms, or the turnover of nutrient energy as it is called in an ecosystem. The rate of turnover usually differs widely from organism to organism. For the same organism it may vary in different situations and at different seasons of the year.

Of the total solar energy that enters a pond, a small portion enters in to the photosynthetic process of green plants and is stored in organic substance (chiefly carbohydrates) in plant bodies. The energy of carbohydrate formed represents the gross plant production or the gross primary production. Actual plant growth is represented by an amount of energy equal to the total carbohydrate formed minus the respiration and is termed as the net primary production. The amount of plant substance produced per unit of time and space is defined as rate of gross primary production, and the rate of net primary production represent the rate at which organic matter is stored.

Comparisons of rates of gross and net primary production in an ecosystem provide interesting data on the utilization of plant food by the plants themselves. In most cases, only a small fraction of net primary production is consumed by organisms at trophic level 2 or by decomposers during a given period of time, the unused portion contributing to the increase of the standing crop. On the contrary, overgrazing of the primary producers decreases the total standing crop. The secondary production constitutes the energy food actually assimilated by heterotrophic organisms, such as primary, secondary and tertiary consumers and the decomposers. The rate of energy storage in these levels is called the rate of secondary production. The rate of production at each higher level decreases from herbivore to top carnivore.

Method for Measuring Productivity

During the last two decades, the interest in the study of productivity of ecosystem has greatly increased and various methods for measuring the rates of primary and secondary production. The basic principles involved in the measurement of productivity, grouped under the headings:

1. Rate of change of standing crop
2. Production of oxygen
3. Uptake of carbon dioxide

4. Concentration of chlorophyll
5. Uptake of nutrients, and
6. Transformation of energy by consumers (secondary production)

1. Rate of Change of Standing Crop

In this method, the growth of macroscopic aquatic vegetation, which develops in a fairly regular seasonal fashion, is periodically measured.

2. Production of Oxygen

In the synthesis of carbohydrate by the process of carbon assimilation, there is a direct quantitative relationship between the amount of carbon dioxide used and the amount of oxygen produced as a by product, as is evident in the expression:

$$CO_2 + H_2O \rightarrow CH_2O + O_2$$

If the organismal respiration is also taken into account, changes in levels of dissolved oxygen reflect gross primary production. The commonly employed method via this idea is the use of 'light and dark bottle' techniques for measuring the rate of change of dissolved oxygen in bottle suspended at depths at which primary production is desired to be known. Since both organismal respiration and photosynthesis co-occur in the 'light' bottle, the 'dark' bottle values for oxygen are used to correct for respiration.

3. Uptake of Carbon Dioxide

This method is based on the assumption that the uptake of carbon dioxide and synthesis of organic carbon are equivalent in the equation mentioned at ii) above and the measurement of carbon dioxide represents production of organic matter.

4. Concentration of Chlorophyll

This method is based on the notion that the chlorophyll content of plants bear a relation to the total organic materials produced by the plants, and estimated chlorophyll concentration would serve and as an index of productivity.

5. Uptake of Nutrients

This method of measuring productivity involves observations .

an absorption of minerals necessary for photosynthesis. The use of radioactive tracers such as C^{14} and P^{32} are modern tools for measurement of productivity.

6. Transformation of Energy by Consumers (Secondary Production)

The measurement of efficiency of energy transformation within the consumer level is very important for the estimation of secondary production. The energy input in the food ingested by a herbivore in its juvenile stage, which precludes its diversion towards development of gonads, is directly used up towards growth and respiration (daily activities). The difference between the energy of ingested substances and that of egested material is assimilated energy which contributes to growth and respiration.

Classification of Water Bodies on the Basis of their Productivity

In India, some hydro biologists have attempted to classify water bodies of known fish yields on the basis of their water quality. Banerjea (1967) surveyed water quality of several ponds. According to him, a neutral condition of *p*H (6.5 to 7.5) is most favorable for productive ponds. Acid water (*p*H 5.5 to 6.5) and highly alkaline waters (*p*H > 8.5) are unproductive and *p*H 7.5 to 8.5 may be considered favorable for average production. Total alkalinity below 20 ppm is indicative of poor production. Under normal condition, concentration of dissolved oxygen above 7 ppm is suitable for a productive water body. Values of dissolved inorganic nitrogen those in the range of 0.1 to 0.2 of average productivity, and those above 0.2 ppm as favorable for productive pond. Dissolved phosphate (as P_2O_5) 0.05 to 0.20 ppm of medium to high productivity and >0.20 ppm of very high fish productivity.

Chapter 17

Management of Freshwater Fish Pond

Scientific management techniques of farm ponds are essential to give maximum yield of fish. Under normal conditions, a pond has a limited capacity to produce fish. However, production can be controlled and increased by various methods, such as correct selection of fish species to be cultured, eradication of weeds and enemies of fish, fertilization and artificial feeding. The object of the fish culturist may be just to produce large quantity of any fish, or to produce good quality of fish of commercial value so as to earn profit. This can be achieved by good management which can be divided into (a) Pre-stocking, (b) Stocking, and (c) Post-stocking management practices.

Prestocking Management of Ponds

Prestocking management practices include the maintenance of pond, removal of aquatic weeds, eradication of undesirable and predatory organisms, improvement of pond water and soil quality and fertilization. Before introducing fry and fingerlings into the pond, it has to be prepared for culture as below:

(*i*) Drying the Pond

If possible, the pond is completely drained of water and dried in sun. This influences the physico-chemical and biological condition of the pond by improving fertility. Fish parasites, their larvae and disease producing organisms are killed. The pond can also be made deeper, if it has become shallow due to deposition of silt.

(*ii*) Control and Eradication of Aquatic Weeds

Aquatic plants growing in the pond are undesirable and harmful for fish culture as they reduce productivity. Aquatic weeds are of the following types:

(*a*) Free Floating Weeds

The leaves float freely on the surface of water and the roots hang inside water. They are abundantly found in wind protected ponds and are harmful, *e.g.*, water hyacinth, *Pistia*, *Lemna*, *Azolla* and *Wolffia*.

(*b*) Emergent Weeds

These are rooted in the bottom but the leaves float on the surface and the flowers rise above the water level, *e.g.*, *Lotus*, *Nymphaea*, *Water lily*, *Nelumbium* and *Nymphoides*.

(*c*) Submerged Weeds

These weeds remain under the water and may or may not be rooted at the bottom, *e.g.*, *Hydrilla*, *Najas*, *Vallisneria*, *Potamogeton*, *Ceratophyllum* and *Utricularia*.

(*d*) Marginal Weeds

These grow along the shallow shore and are rooted, *e.g.*, *Typha*, *Ipomea* and *Sagittaria*.

Several species of filamentous algae form thick mats or scums on the water, or lie scattered. These are *Spirogyra*, *Oscillatoria*, *Anabaena* and *Microcystis*. Aquatic plants in limited quantity are useful and necessary part of the ecology of the pond as they form natural food for many herbivorous fishes. They oxygenate water fertilize the pond on decay, and reduce turbidity. They are used as shelter by some fish and form spawning beds. But excessive growth of weeds is harmful to the fish, as they obstruct free movement of the fish, and cause oxygen depletion and accumulation of carbon dioxide. They consume nutrients of the pond, obstruct netting of fish, and cause silting of the pond

Control of Weeds

Three methods *i.e.* mechanical, chemical and biological are used to eradicate weeds from ponds. Removal of weeds by hand picking, uprooting and cutting using manual labour is quite satisfactory in small bodies of water. Rooted submerged weeds are removed by using log weeders fitted with spikes and barbed wire. Several common floating weeds like the *Pistia*, *Salvinia*, *Azolla* and *Lemna* are removed completely by manual labours, dried and burnt for eradication. They are also used for preparing good compost manure.

In large bodies of water, when the vegetation is dense, various chemicals or weedicides are used for controlling weeds. The chemical to be used should be cheap and easily available, and non-toxic to fish and man. Weedicides commonly used are 2,4-D (2,4-dichlorophenoxy acetic acid) also known as Taficide, Hexamar, Fernoxone. Other chemicals as Chloroxone, Simazine, Methoxane, Agroxane, Copper sulphate, Calcium cyanide, Sodium chloride, Sodium arsenite, Diuron, etc. are also used to eradicate weeds effectively.

Sometimes filamentous algae as the *Spirogyra*, *Pythophora* grow extensively forming thick mats. *Anabaena*, *Microcystis* and *Oscillatoria* also grow extensively and form algal blooms, which are harmful to fish and kill them by chocking the gills, causing oxygen depletion or by rotting. Chemicals, which can be successfully used to control different types of weeds are given below (Table 17.1).

Several types of weeds can be easily controlled by introducing a suitable herbivorous species of fish, which feed on them. Useful are the grass carp, common carp, tilapia and gourami. The grass carp, *Ctenopharyngodon idella*, is most useful species and feeds on submerged weeds like the *Hydrilla*, *Najas*, *Ceratophyllum*, *Vallisneria* and *Potamogeton*. Advanced fry and fingerlings of grass carp consume *Wolffia*, *Lemna*, *Azolla* and *Salvinia*. The common carp, *Cyprinus carpio* is also useful in keeping the weeds in check. Besides fishes other animals like ducks can also be used for biological control of weeds. Biological method using fish and duck is probably the best method for controlling weeds.

(*iii*) Preparation of the Nursery Ponds

Fry and hatchlings are stocked in small, shallow ponds called nurseries (1.0 to 1.5 m deep), while fingerlings are stocked in deeper ponds (1.8-2.5 m deep) for rearing. The nursery, rearing and stocking ponds need careful preparation before introducing fry or fingerlings, so as to reduce mortality. Mortality is due to the following factors:

1. Predatory and weed fishes are present in the pond.
2. Predatory aquatic insects may be present.
3. Suitable food is not available to the fry.
4. Fry and fingerlings of different size and age, if stocked in the same pond, exhibit cannibalism.
5. Over growth of phytoplankton causes depletion of oxygen.
6. Large differences in the temperature and pH of the water in hapa and pond, may cause mortality.

Small sized ponds can be drained of water and dried in the sun to kill harmful organisms and all enemies of fish. The ponds are then filled with freshwater.

(*iv*) Eradication of Predatory and Weed Fishes

Predatory fishes if present in ponds are harmful as they eat fry and fingerlings. Minor carps which do not grow to large size are undesirable in the ponds and are called "weed fishes". They multiply rapidly and consume large amount of food, and thus compete with the fry of major carps for food. Common predatory and weed fishes found in ponds are:

Predatory Fishes

Wallago attu, *Channa* spp., *Notopterus chitala*, *N. notopterus*, *Mystus* spp., *Clarias batrachus*, *Ompok pabdo*, *O. bimaculatus*, *Silonia silondia*, *Mastacembelus armatus*, etc.

Weed Fishes

Puntius ticto, *P. sophore*, *Esomus danricus*, *Oxygaster bacaila*, *Amblypharyngodon mola*, *Ambassis nama*, *A. ranga*, *Barilius* spp., *Gadusia chapra*, *Colisa fasciatus*, *Xenentodon cancilla*.

Predatory and weed fishes can be removed from the pond by repeated netting, but this does not ensure complete eradication, as more species such as the murrels, mangur, singhi, etc. burrow into the mud, and escape netting. Another method is to drain the pond completely and dry it in the sun. However, this is not always possible, or is uneconomical. Hence, fish poisons are used to eradicate predatory and weed fishes.

Fish Poisons

Unwanted fish can be easily killed by using a piscicide. Derris powder is a plant derivative containing 5 per cent rotenone ($C_{23}H_{22}O_6$) and is commonly used in ponds. It is a contact poison and the fishes are killed due to injury to the gill membrane. A dose of 6-10 ppm is sufficient to paralyse all species of fish including the air breathing fishes. The required chemical is dissolved in water and sprayed on the pond surface, and then splashed for thorough mixing. This poison is very effective on sunny days and all fish are killed within a few hours, which are fit for human consumption after 24 hours. The effect of the poison lasts for about a week. Derris powder is used about a month before stocking, so that enough time is available for manuring the pond and growth of zooplanktons.

Some other plant derivatives can be used if derris powder is not available. These are prepared from Akhrot (*Juglans* spp.), Kuchla (*Strychnos* spp.), Banlu (*Dioscorea* spp.), and Ban Tambaku. Powdered seed of *Croton* and *Barringtonia* are also effective as fish poison.

Mahua oil cake (*Bassia latifolia*) is toxic to the fish and contains about 6 per cent saponin which enters the blood stream through gills and haemolyses red blood cells causing death of fish. Mahua oil cake at 200-250 ppm kills all predatory and weed fishes, and later serves as a manure. The poisonous effects lasts for about 3 weeks.

Certain chemicals as Aldrin, Dieldrin and Endrin are toxic and kill fishes, and the toxicity lasts for 2-3 weeks. Effective dose of endrin is 0.01 ppm, as it is most poisonous. Tafdrin-2p, contains 20 per cent endrin and is recommended to be used at 0.01 ppm. It is dissolved in water and the solution is sprayed over the surface of the pond. Water is mixed thoroughly and the fish die within 2-4 hrs. It should be used 5-6 weeks before stocking to ensure detoxification of water. Certain organophosphates as the phosphamidon, Dimethyl-dichlorovinyl phosphate (DDVP.) are also useful in killing and eradicating predatory and weed fishes. However, use of chemicals should be avoided, and derivatives like the derris powder should be preferred.

(*v*) Control of Aquatic Insects

A number of predatory insects as the backswimmers, giant water bug (*Belostoma*), water scorpion (*Nepa*) and beetles enter the nurseries, especially during monsoon and prey on spawn and fry. All these species of insects and nymphs of dragon fly are highly predatory, eating fry and causing heavy mortality.

Insects can be controlled by frequent netting with small meshed nets, but many insects escape netting and fly back. They can be eliminated by using insecticides, but these chemicals kill zooplankton and fish spawn and fry also, and the use of insecticides is not practicable. Application of a mixture of soap and oil is the most effective method for controlling aquatic insects. Soap is dissolved boiling water and mixed with mustard or coconut oil. The mixture is sprayed over the surface of water on a windless day. 8 kg of cheap soap dissolved in hot water mixed with 25 kg of oil is enough for one acre of water. The oily film formed on the water chokes the respiratory tube of insects. The effect of oily film lasts for a few hours and all aquatic insects are killed. It should used one day before the introduction of fry, and the operation can be repeated later also as the emulsion does not harm fry and plankton.

(*vi*) Fertilization of the Pond

The natural productivity of the pond can be increased by using organic or inorganic fertilizers which provide nutrients, vitamins and minerals.

Liming the Pond

The first step in fertilization of a pond is the application of quick lime (CaO) which rises the pH of water and is antiparasitic killing bacteria and parasites and their larval stages. Quick lime @ 200 kg/ha is generally spread on the bottom, 15 days before stocking the pond with fish fry and fingerlings. For getting full benefit, the pond should be left to dry for two weeks after liming. Liming should be done if the water is acidic (pH 4.5-6.5), soil is too muddy and organic matter high. Presence of calcium speeds up decomposition of the organic matter releasing CO_2 from the bottom. Lime can be applied @ 300-500 kg/ha depending on the pH of the soil.

Manuring

Natural food production of the pond has to be enhanced to support a large number of fry stocked in it. Both inorganic and organic fertilizers are used to increase the available food. Inorganic nitrogenous fertilizers used are sodium nitrate, ammonium sulphate, ammonium nitrate, ammonium carbonate, urea, etc. super phosphates are also widely used. Ponds are treated with fertilizers after liming, but these are costly, and bring about phytoplankton bloom, and heavy growth of zooplankton also. The use of inorganic fertilizers depends on the available nitrogen and phosphorus in the soil.

Cow dung as an organic fertilizer is preferred as it supplies all the nutrients needed by the fish. A dose of 5000-10000 kg/ha cow dung causes the production of thick swarms of zooplankton (rotifers, copepods, cladocera) in about 15 days. Manuring should be done 10 days after poisoning with fish toxicants. Cow dung is sprayed on the bottom or placed in heaps along the bank of the pond. It is considered better to apply cow dung in two doses instead of one. Good quantity of zooplankton can be maintained for a longer period, if a dose of 10000 kg/ha cow dung is used two weeks before stocking, and a second installment of 5000 kg/ha is given one week after stocking.

Liquid manure from stables, guano, farm manure consisting of cow and pig dung are extensively used for manuring and bring about heavy growth of zoo- and phytoplankton. Sewage is also good fertilizer and enhances the growth rate of fish. A mixture of superphosphate of lime, cow dung and mahua oil cake has been successfully used as a manure are soybean meal, cotton seed meal, mustard oil cake etc. It has been suggested that the inorganic fertilizers should be used in ten equal monthly installments, alternately with organic and inorganic manure.

Sometimes, use of organic and inorganic fertilizers results in excessive growth of algae and phytoplankton. If the water turns green due to algal bloom, cow dung is applied in the form of liquid spray, so that the water becomes turbid brown. This checks penetration of light, and the algae die in 4-5 days.

Stocking

Stocking phase consists evaluation of carrying capacity of the pond, relative density and combination of selected species.

High production of fish from a pond is possible if a desirable species is selected for culture considering the type of pond, its ecological conditions and facilities for supplementary food. The fry should be introduced into the pond at a suitable time when the concentration of zooplankton is maximum. The stocking rate varies from 10-20 lakh fry per ha, depending upon the density of zooplankton and the carrying capacity of the pond, which is related to the natural production of food, manuring, feeding etc. If the nurseries are well manured and supplementary food is given, 50-60 per cent of fry survive and grow to fingerlings in about 15 days. Thus, two or three crops can be raised in one season.

Choice of a suitable species is an important factor in achieving high yield. As a single species can not utilize all available food in the pond, a proper combination of 3-5 species of carps gives best results. These should be fast growing compatible species which occupy different ecological niches. This method is called 'polyculture' or mixed farming. Generally, Indian major carps, catla, rohu and mrigala are grown together in a pond and give a high yield.

Poststocking Management

Post-stocking management practices mainly comprise the supplementary feeding, fish health monitoring, sampling and harvesting pattern of the fish crop.

(*i*) Feeding

Fry feed voraciously on the zooplankton of the pond and the available food is consumed to a large extent within 4-5 days. Hence, to ensure proper growth of fish, supplementary feeding is necessary so that high density of fish can be maintained. Natural food production in the pond is limited but it provides all the essential nutrients to the fish, and is rich in protein. It consists of zoo- and phytoplankton, crustaceans, insects, their larvae, worms, mollusks, etc. Use of inorganic and organic fertilizers increases the natural food, but still is not enough to sustain a large population of fry for long time. Hence, a mixture of natural and artificial food gives best results.

The artificial food should be such that it is acceptable to the fish, is easily digestible, cheap and readily available in large quantity. It should have good nutritive value and high conversion rate. Generally, artificial food is of vegetable origin and consists of rice, barley, wheat and pulses, leaves, tuber roots, oil cake, kitchen waste are also used as food. Finely powdered rice bran, paddy powder, and oil cakes of coconut, mustard and groundnut are used as food for fishes. The food is finely ground and sieved before feeding the fry. Fry Indian major carps, a mixture of rice bran and mustard oil cake or groundnut oil cake (1:1 by weight) is considered to be the best.

Table 17.1: Weeds and their Control

Weedicide	*Weeds Killed*	*Dose Used*
2, 4-D	Water hyacinth, *Eichornia*, *Lemna*, *Pistia*, *Azolla*	4.5-6.5 kg/ha
Taficide-80 (2,4-D Sodium salt)	*Pistia*, *Lemna Typha*, *Colocasia*, *Nelumbo*, *Nymphoides*	1–1.5% solution in combination with 0.25% soln. of "surf"
Sodium arsenite	*Lemna* and *Pistia*	1% solution
Copper sulphate alone or Ammonium salt	Submerged weeds like *Hydrilla*, *Vallisneria*, *Potamogeton*, *Naja* *Ceratophyllum*, algal bloom	175 kg/ha
Simazine	Algal bloom	0.5-1.0 ppm
Diuron	Algal bloom	0.1-0.3 ppm
Superphosphate	*Najas*, *Potamogeton*, *Chara*	500 ppm
Chloroxone	Floating weeds	3.5 kg in 500 lit. of water

Certain species (trout, salmon) prefer animal food, which consists of fresh or dried fish, fish meal, meat, fresh or dried shrimps, waste from slaughter houses (liver, lung, spleen, blood etc), silk worm pupae, larvae insects, snails, worms, etc.

Artificial food is given daily @ 2-3 per cent of the fish biomass. It is either sprayed over the water surface or kept as thick paste in shallow, wide mouth earthen pots or trays at a few spots in the pond. Feeding is usually done in the morning or evening, and should not change suddenly.

Thinning and Harvesting

Advanced fry of about 3-4 cm in size are netted and released in rearing ponds from time to time. Rearing ponds are larger than nursery ponds are have to be prepared in the same way as to eradicate aquatic weeds and predatory fishes. Food requirements of advanced fry are different than the young fry. They are given artificial food at regular intervals for maintaining fast growth. Proper management of the rearing ponds results in 70-80 per cent survival of carp fry.

In about two months fry attain a size of 8-13 cm and are transferred to separate stocking ponds which are of large size. The stocking ponds are also treated in the same way to clear aquatic weeds and undesirable fishes, and are manured to enhance production of fish food. A mixture of inorganic fertilizers (ammonium sulphate, super phosphate and ammonium nitrate) @ 1000-1200 kg/ha/yr, and cow dung @ 20000 kg/ha/yr, is recommended for stocking ponds.

For best utilization of different types of food, a combination of catla, rohu and mrigala in the ratio of 3 : 3 : 4 is cultured in a pond. Silver carp and grass carp are also now included for mixed farming. Excessive growth of algae must be checked because they restrict the movement of the fish, cause oxygen depletion, choke the gills, and pollute water by rotting. The process of thinning, stocking, manuring and artificial feeding is carried out from time to time. Large sized full grown fish (25-35 cm or more) are harvested and transported to the market. The measures as described above, if adopted carefully, result in a high yield of good quality fish.

Chapter 18

Economics of Composite Fish Culture

Fish production by culturing a single species in a pond using old traditional methods gives a low production. "In order to obtain high production per ha of water body, fast growing compatible species of fish of different feeding habits, are stocked together in the same pond so that all its ecological niches are occupied by fishes. This system of pond management is called mixed fish farming or composite fish culture or polyculture.

The basic principle of composite fish culture is that when compatible fishes of different feeding habits are stocked together, they secure for themselves in the most efficient manner, all life requisites available in the pond for fish production without harming each other. In fact, there does not obtain any serious competition between different species but each species may have a beneficial influence on growth and production of the other. For example, the grass carp, by consuming aquatic vegetation, convert plant tissue in to fish flesh but its excreta fertilizes the pond which benefits all other species.

Mixed fish farming first started in ancient China. In India, mixed fish culture is an age-old practice. In Bengal, catla, rohu and mrigal in the ratio 3:3:4 are reported to be commonly stocked in ponds (Alikunhi, 1957).

Hora and Pillay (1962) have however, recommended the following stocking proportion:

Catla	30	30
Rohu	60	50
Mrigal	10	10
Calbasu	×	10

Stocking of fingerlings (8-13 cm long) per ha:

Catla	1,875	1,875
Rohu	3,750	3,125
Mrigal	625	625
Calbasu	–	625

Indian major carps, catla, rohu, and mrigal are generally stocked together, and sometimes calbasu is also added to this group, and they are stocked in the ratio of 3:3:3:1. The Indian major carps can be cultured in the ratio, *Catla catla* 40 per cent; *Labeo rohita* 30 per cent; *Cirrhinus mrigala* 20 per cent; and *Labeo calbasu* 10 per cent.

Composite farming under three broad combination *viz.*

1. Culture of Indian major carps alone
2. Culture of exotic carps alone
3. Culture of Indian major carps and exotic carps.

Culture a single species is not profitable. Therefore a group of fishes are selected each having different feeding habits from others so that all the food available in different zones of the pond is used profitably. During the recent years the production of fish can be increased by 8 or 9 times, if the following indigenous and exotic species are cultured together: *Catla catla; Labeo rohita; Cirrhina mrigala; Hypopthalmichthys molitrix; Ctenopharyngodon idella; Cyprinus carpio.*

Catla is the surface feeder and feeds mainly on the zooplankton and detritus. Mrigal and Calbasu, bottom feeder-feeds on decayed vegetation benthic animals, plants and epiphytic plankton from the bottom of the pond. Rohu column feeder, feeds on decayed vegetable matter, epiphytes etc. can also be utilized. Silver carp surface feeder, feeds only on phytoplankton unlike catla which feeds mainly on zooplankton. Common carp is an omnivorous bottom feeder utilizing mainly the food which *C. mrigala* is not able to consume. Grass carp is known for its efficient utilization of all noxious and excessive macro-vegetation and occupies on ecological niche not utilized by

Table 18.1: Economics of Composite Fish Culture for One Year (Anon., 1969 and Lakshmanan *et al.*, 1971)

	Indian Major Carps (Catla, Rohu and Mrigal)	*Exotic Carps (Silver, Grass and C. carp)*	*Indian Major Carps and Exotic Carps*
Pond size (ha)	0.12	0.12	0.13
Stocking density	3750/450	3700/450	5075/666
No. of fingerlings/ha/pond			
Economics calculated (calculated per ha/yr)			
(*i*) Income (Rs.)	8,772.00 (A)	10,860.00 (A)	15,787.50 (A)
(*ii*) Expenditure (Rs.)			
a) Pond preparation	100.00	100.00	100.00
b) Fingerlings	375.00	370.00	500.00
c) Fertilizers	855.00	553.00	918.75
d) Artificial feed	612.50	300.00	625.00
e) Labour	375.00	375.00	980.00
f) Nets	500.00	500.00	500.00
Total	2,817.14 (B)	2,198.00 (B)	3,623.75 (A)
(*iii*) Net profit	**5,954. 26** (A-B)	**8,662.00** (A-B)	**12,163.75** (A–B)

only other fish in this association. Rather its faecal matter further serves as food for the common carp and also acceleration plankton production and growth of the fishes.

Experiments have shown that Indian major carps when stocked along with three species of exotic carps, the yield goes up 8-9 times, as compared to monoculture. These exotic carps are silver carp, grass carp and common carp. Thus, these six species have complementary feeding habits and do not harm each other, and constitute a good combination for polyculture. For maximum utilization of all the food resources of the pond, these species are stocked at a rate of 8000-8500 fingerlings per ha, and the ratio recommended is 40 per cent surface feeders, 20 per cent column feeders, 30 per cent bottom feeders and 10-20 per cent macro-vegetation feeder.

The stocking rate and proportion of various species varies depending on the size of the pond, availability of various kinds of fish food and the physico-chemical condition of the water. With proper fertilization and manuring of the pond, it is possible to achieve an yield up to 8000 kg per ha in one year.

Jhingran, 1974, found that, 1) Average yield of indigenous major carps in intensive mixed farming is about 4000 kg/ha/yr; 2) That of exotic carps is about 3000 kg/ha/yr; and 3) The average yield of indigenous and exotic species cultivated together is 8000 kg/ha/yr. The ratio used as follows:

1. 1:1:1:2:5:2:5:2;
2. 1:1:2:5:1:5:3; and
3. 1:5:3:1:25:1:2, for the six species in order given above.

Anjana fish farm, west Bengal, an average production of 4000 kg/ha/yr has been obtained by composite fish culture. In experiments conducted at the pond culture division of the CIFRI at Cuttack a record yield of more than 9000 kg/ha/yr was obtained. The following six species were used in the ratio mentioned below:

Grass carp	25
Silver carp	15
Catla	15
Rohu	30
Scale carp	40
Mirror carp (common carp)	10

Chapter 19
Culture of Exotic Fishes

Several species of good quality fish have been imported into India and are cultured. They are called exotic species and become adopted to the new environmental conditions. They belong to three groups: Game fishes, Food fishes, and Larvicidal fishes.

Important exotic fishes introduced in India for sport, food and mosquito control.

Species	*Source*
Food Fishes	
Ctenopharyngodon idella	Japan
Hypopthalmichthys molitrix	Hong Kong
Cyprinus carpio	Bangkok
Osphronemus goramy	Java and Mauritius
Tilapia mossambica	Bangkok
Carassius carassius	England
Game Fishes	
Salmo irrideus (Rainbow trout)	Europe
Salmo trutta (Brown trout)	England
Salmo gairdneri	Ceylon
Oncorhynchus nerka	Japan
Larvicidal Fishes	
Gambusia affinis	Italy
Lebistes reticulates	South America

The following are the more important from the viewpoint of fish culture. A description of the bionomics, breeding and cultural techniques of these species is given below.

Common Carp (*Cyprinus carpio*)

It is commonly called the mirror carp or the common carp. Three varieties of this fish are known; scale carp (*Cyprinus carpio communis*), mirror carp (*Cyprinus carpio specularis*) and leather carp (*Cyprinus carpio nudus*).

It is world wide most extensively cultivate species. In India it is cultivated singly as well as along with Indian major carps. Voraciously omnivorous, grow very fast and is prone to artificial feeds. It is non-predatory. When cultured with Indian major carps it competes for food mostly with mrigal and calbasu. The strain of scale carp obtained at Cuttack from Bangkok in 1957. In the plains of India, the fish attains maturity within 6 months of hatching, where as in the upland lakes it takes almost a year to mature. Common carp naturally breed in confined waters. Spawning occurs in shallow marginal, weed-infested areas. Fertilized eggs of common carp are small, spherical, demersal, and adhesive. Embryos hatch out within 71/2 days at the water temperature of 12°C; 31/2 days temperature 20°C and within 2 days at 28°C to 32°C. The newly hatched out larvae adheres to the leaves of aquatic plants by means of cement glands till yolk is partially absorbed. The yolk is absorbed within 2 to 6 days depending upon temp.

Breeding Technique

Common carp breed throughout the year in tropical climate with two peak breeding periods one lasting from mid-January to March and the other during July and August. Males and female breeders are segregated a few months prior to the breeding season and stocked in separate ponds preferably free from other fishes. The distinguishing characters of male and female spawners:

Male: Vent not projecting and deep pit like; when pressed belly gently oozes milt; pectoral fin rough only in some ripe males.

Female: Vent slightly projecting like a narrow median silt; pectoral fins smooth.

Breeders are regularly fed on artificial feed comprising mixture of oilcake and rice bran in 1:1 ratio at 2-3 per cent of their body weight. Fully matured breeders are used for reproduction. A mature female possesses an almost rounded soft and bulging abdomen and vent projecting into a small papilla-like structure. On male when pressed gently on their abdomen, release milt through the vent.

Breeding is done either in 'hapas' or cement cisterns or in the pond itself. For this purpose, a set of brood fish comprising 1 female and 2 or 3 males are introduced inside a breeding hapa keeping the total weight of male nearly equal to the weight of the females.

The breeding hapa is almost of the same type as used for major carps except that the top flap is unstitched on three sides. Common submerged aquatic weeds like *Hydrilla* and *Najas* are generally used as egg collections. The aquatic plants use as egg collectors are thoroughly washed before hand. In normal practice, about 2 kg of weeds are considered adequate per kg weight of female but, for larger fish weighing 3 to 5 kg, about 4 to 7 kg of egg collectors are required.

The best time for release brood fish in the hapa is evening. Fishes generally spawn in installments within 6 to 10 hours after being released in the hapa, and by morning, the egg collectors, with million of eggs gently transferred to hatching hapas. After spawning is over the brood fish are removed from the breeding hapas and weighed. The difference in weight before and after spawning gives the weight of eggs released.

12 to 15 per cent is given for the faecal droppings and the total number of eggs released is calculated at the rate of about 700 per gram differences in weight of fish (Alikunhi, 1966). About 1 kg of the egg collector carrying nearly 40,000 to 1,00000 eggs can be put in a single hatching hapa. Four to five days after breeding the spawn is collected from the hatching hapa, care being taken to remove the weeds first.

Spawn from the hatching hapa is sieved through a coarse mosquito netting cloth to remove debris and permit its measurement. For measurement spawn is collected over a piece of cotton cloth and scooped into a small aluminum cup of known capacity. The number of spawn per cup are determined at the rate of 400 spawn per ml. The spawn is then stocked in pre-prepared nurseries.

Culture of Common Carp

Common carp is generally cultured in combination with Indian carps. The technique for the preparation of nurseries to rear the spawn of common carp is the same as that adopted for major carps. According to Alikunhi (1966) common carp nurseries and rearing ponds, are heavily manured with cowdung at the rate of 2,500 to 5,000 kg/ha. Stocking density 2,500/ha. Growth average 600 to 800 g and 1 kg, respectively in the first year. 1,500 to 2000 g in the second year (stocking density 1,500/ha).

Silver Carp (*Hypophthalmichthys molitrix*) (Valenciennes)

It is commonly known as 'silver carp'. It is native of China. In India, fingerlings of silver carps was brought from Japan, to the pond culture division of CIFRI, Cuttack, Orissa. They were breed successfully by hypophysation and their fry distributed from Cuttack to various states of India. Silver carp feed on unicellular algae, zooplanktons, decayed macro vegetation and detritus. Attains maturity 2 years old. Silver carps bred first time successfully by hypophysation during June-July, 1962 (Alikunhi *et al.*, 1963). The technique of breeding of Chinese carps in India is the same as that for Indian major carps. Key identification characters of brooders:

Male: Pectoral fins rough and freely oozing milt when the abdomen is gently pressed.

Female: Soft, bulging abdomen, pinkish and slightly protruding vent are selected.

One set of breeders consists of 1 female and 2 males. The female is given usually two injections and males one at a time of the second injection to the female. The female start oozing 6 to 8 hours after the second injection. The eggs are striped and fertilized by dry method. Synahorin along with pituitary (10 and 11 mg/kg doses) gave positive results.

Culture

Nursery ponds for rearing spawn of silver carp are prepared as same like Indian major carps. At a stocking rate spawn of 5 lakhs per ha (average length 7.4 mm) reared for 9 days. Fry (22-24 mm in size) were released in rearing ponds at the rate of 6,250 to 25,000 per ha for 10 days attain length 48.3 mm and average weight 1.30 gm. Silver carp fingerlings (average weight 64 g) stocked in pond at the rate of 4000/ha attained average weight of 700 g and gross/net production 1558/1374 kg/ha/yr (Barrackpore, 1970). Silver carp being a plankton feeder has been reported by Alikunhi and Sukumaran (1964) to compete for food with catla.

In 1962, when silver carp and catla fingerlings in the ratio 1:1 were stocked @ 500/ha (wt of stocked fish 1.5 kg) for 15 to 36 days. The production of both the species was computed to be 900 kg/ha/ year. Silver carp showed faster growth than the catla and accounted for 61 per cent of the production (Alikunhi *et al.*, 1971). The ratio of 2:1 between silver carp and catla gave maximum production.

Grass Carp (*Ctenopharyngodon idella*) (Valenciennes)

It is commonly called as grass carp. Is the natural inhabitant of the flat land rivers of China. It is also found in Russia. The first fingerlings of grass carp were brought from Hong Kong to India 1959, Central Inland Fisheries Research Institute (CIFRI), Cuttack, Orissa. They were successfully breed by hypophysation at Cuttack in 1962. Male mature when 2 years old and females matured when 3 years old. But offspring's obtained as a result of hypophysation in India, matured within 2 years in ponds at Cuttack. Its growth is rapid depending upon the availability of suitable weeds. According to Alikunhi and Sukumaran (1964) grass carp attained a weight of 4.5–7 kg when 3 years old, in pond at Cuttack. The fry of grass carp feed on zooplanktons and phytoplanktons, but the adult is a herbivorous fish.

Culture

Spawn stocked in a nursery pond @ 1,25,000/ha attained 38 to 52 mm in length and an average weight of 1.5 g during 15 days (Alikunhi *et al.*, 1963).

Alikunhi and Sukumaran (1964) observed that grass carp of an average size of 26.5 cm and weight 215 g when stocked @ 125 fish per ha in a pond containing a thick growth of *Hydrilla*, measured in 3 weeks on an average 40.3 cm and weighed 837 g each.

Chakrabarti *et al.* (1979) used 184 grass carp yearlings (800 g each) in a pond containing dense *Hydrilla*, *Najas* and *Potamogeton* in large water bodies about a 2 and 10 ha in area. Cleared an infestation of 85.8 tonnes of weeds in a 1.8 ha pond in 103 days. Survival rate 98.4 per cent average weight of 3,045 g.

Tilapia (*Tilapia mossambica*)

It is commonly called as tilapia. The natural habitat of Tilapia in rivers of the East coast of Africa. In India, the first time tilapia was brought by the Central Marine Fisheries Research Institute, Mandapum on August, 7, 1952, from Bangkok, and the second by the Madras fisheries Dept. in the same year from Ceylon.

Tilapia is a herbivore fry feed exclusively on diatoms and other unicellular planktonic and epiphytic algae. The adult tilapia feed mainly on vegetable food. When vegetable food are scarce, worms, insects, crustaceans, fish larvae are eaten.

According to Vaas and Hofstede (1952) tilapia its first maturity when about 2 months old and 9-10 cm long. Tilapia breeds through the year at an interval of 15–25 days. The frequency of spawning varies according as the climate and food. Spawning occurs during the warm season when temperature reaches 20°C. The fecundity of tilapia is between 80 and 1000 eggs. The male prepares a shallow circular nest in the sand and the eggs are laid by the female after brief courtship. The eggs are fertilized by the male, and soon the female picks up the eggs in her mouth, thus exhibiting buccal incubation. Hatching takes palce in 3–4 days, and 8.0 mm fry are liberated from the mouth. It can tolerate wide range of salinity and is therefore, recommended for culture in coastal lagoons.

Culture

Tilapia is a fast growing hardy fish, but does not survive at temperature lower than 10°C. Tilapia can tolerate a wide range of salinity change in salinity from freshwater to saline waters of salinity 30 to 48 per cent (Panikkar and Tampi, 1954). Sudden change in salinity proves fatal to the fish. Tilapia can culture also coastal lagoons. Vaas and Hoftede (1952) reported that tilapia grow fast, tilapia young weighing 3-5 g grew to an average weight of 15 to 25 g in 40 days. According to Hora and Pillay (1962), under favorable ecological conditions, tilapia could attain a weight of 850 g at the end of the first year. An extensive culture of tilapia the pond was regularly manured with cowdung and N: P: K fertilizers and the fish stocked were fed with mustard oil cake. Tilapia when cultured with major carp in ponds, was observed to feed on carp spawn even in presence of its natural food items.

In Indonesia, tilapia is cultured in sewage water mixed with freshwater and given an yield of 7000 kg/ha/yr. It can be easily cultured in paddy fields also, it is useful as a forage fish, and is cultured along with carnivorous and piscivorous species as Channa spp., and serves as their food.

For natural breeding, the male fish makes a shallow circular pit 25 cm in diameter and 6 cm deep by scooping out mouthfuls of sand and depositing it nears the margin of pit. During the process of courtship, the female extrudes eggs in one batch only. Then the male passes-nest rubbing its genital tube over the newly eggs. The eggs hatch in 3 to 5 days.

Chapter 20

Methods of Pisciculture

Pen Culture

This is simply an enclosure, often along banks and shores, of some device which utilizes the bank or the shore line as one of its sides. Fishes are then raised in this enclosure. Pens are made of wood or bamboo and are set afloat in rivers. They are covered with straw mats or aquatic plants to protect the cage fish from the sun.

Pens are of 3 sizes:

1. *Small*: 1 m x 2 m x 1 m deep, used mainly for rearing fry
2. *Medium*: 2 m x 5 m x 1.5 m deep, and
3. *Large*: 2.5 to 4.5 m x 7 to 10 m x 1.5 to 2 m deep.

The stocking rate is 2000 fry for smaller pens, 1,000 fingerlings and juveniles for medium-sized pens and 1,500 to 3,000 fingerlings and juveniles for large pens. In Cambodia, juvenile fish, each weighing 80 to 150 g, are preferred by fish farmers for stocking in pens. *Pangasius pangasius* is generally cultured in pen. *Pangasius* are fed two or three times a day, either by throwing the feed into the pen or by placing it in a feeding tray hung submerged inside the pen. The amount of feed given daily ranges between 4 to 6 per cent of the weight of the fish.

Pangasius pangasius is a hardy fish which feeds on offal, gastropods, insects, plankton, etc. The culture of *Pangasius pangasius* as a food fish has been reported in Khulna district of East Bengal (now Bangla Desh) (Hora, 1952). Ling (1967) has described *Pangasius* culture both Thiland and Cambodia in pen. In Japan, tuna (*Thunnus thynnus*) cultured in pen.

Small Pangasius each weighing 80 to 100 g at the time of stocking may grow to an average weight of 1 to 1.2 kg in 8 to 10 months.

Cage Culture

Cage culture of various species of fish, both freshwater and marine, is now practiced in several countries like Japan, Cambodia, USA, and Indonesia. In India also cage culture has been introduced in Assam, Bihar, Karnataka and some other states. Cage culture gives much higher yield as compared to pond culture; and the technique has been applied successfully for culturing carps, càtfishes, air-breathing fishes, and clupeids. Carnivorous species and those inhabiting bottom mud can also be cultured in cages.

The species which are generally cultured in cage in various countries are: *Cyprinus carpio* (common carp), *Barbus* spp., *Tilapia, Pangasius, Clarias batrachus, Channa* spp., *Mystus* spp., *Salmon* spp. (trout), *Oncorhynchus, Clupeid yellow tail, Notopterus* and *Anabas*.

Even a small perennial pond is sufficient for this purpose, and several cages can be used in one pond. Swamps and muddy ponds infested with weeds, have been profitably used for cage culture.

There is no fixed design or size of the cage for culture. They are made of galvanized welded wires and nylon meshes in USA, Japan, etc. In India, a cage is made of split bamboo and are of various size, ranging from a few square meters to 12-15 sq meters. For carp culture a cage is generally 2m deep. The shape is usually rectangular, and a cage may be stationary fixed to some strong poles or free floating for which steel drums or synthetic floats are used. The size of the mesh is small in cages in which fry are to be cultured. These are called "rearing cages". However, as the fry grow to fingerlings, they are transferred to stocking cages, which have large sized mesh about 4 cm to allow free circulation of water through the cages. The number of fingerlings to be stocked depends upon the size, depth and surface area of the cage and the species of fish to be cultured. In Japan, the standard size of a cage is 10-12 sq meters, and fifty-fish/sq m are stocked in it.

Feeding

Due to high density of stocking, and the fish being confined in cages, supply of artificial food is necessary. The food is supplied in the form of pellets composed of about 70 per cent animal and 30 per cent plant material. The raw material consists of rice bran, soybean, oil-cake, meat, liver, blood and fish. The quantity of food supplied varies with the species and number stocked in a particular cage. Feeding has to be so regulated that all food is consumed before being washed out with water. In the case of carps, feeding may be done 4-6 times per day.

The growth of fish is very fast inside the cage. For example in Japan, the yield of yellow tails was 30 kg/sq.m/yr, and for carps 4000 kg/ha. In experiments conducted in Assam, air breathing fishes like the murrels were grown in cages in weed infested water, and the yield was 25,000–45,000 kg/ha in 200 days. Mixed culture of compatible species (*Clarias*, *Anabas* and *Heteropneustes*) has also been

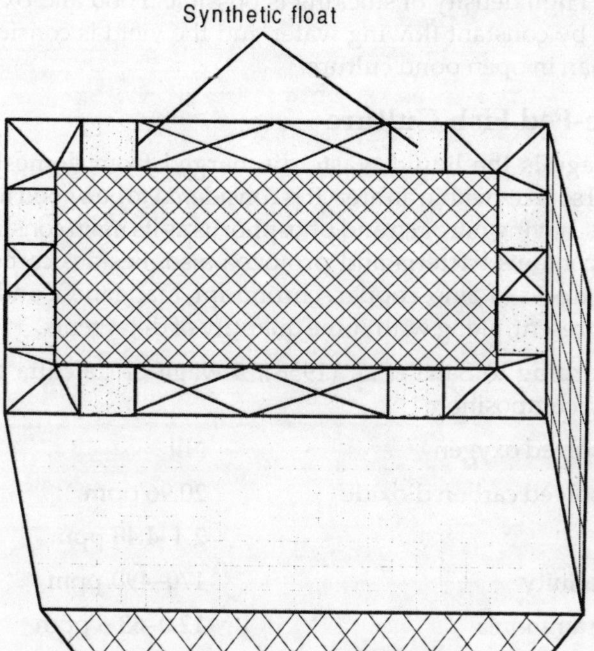

Figure 20.1: A Floating Cage

successful in swampy waters. In Bihar, Assam, Madhya Pradesh and Uttar Pradesh, several ponds are used for raising cash crop of "makhana". (*Euryale ferox*), and "singhara" (*Trapa bispinosa*). These waters can also be used for the culture of air breathing and other fishes in cages, increasing the income of the farmer.

Pangasius is a hardy, carnivorous fish feeding on gastropods, mussels, insects and is easily cultured in cages and pens made of wood or bamboo. The stocking rate is 1000 fingerlings for a cage of medium size and 1500-3000 in large ones. Artificial food consisting of chopped trash fish, banana and rice bran is given @ 4-6 per cent of the weight of the fish per day. The fish grows to 1- 4 kg in 8-10 months.

Thus, cage culture is an efficient and effective method of utilizing swampy and weedy water. It requires little investment. Harvesting is very easy. Only a small part of the pond is used for cages, and the rest can be used in the normal way. Culture of carnivorous and predatory fishes is very convenient, young fry remain protected from enemies. High density of stocking is possible. Food and oxygen is supplied by constant flowing water and the yield is considerably higher than in open pond culture.

Sewage-Fed Fish Culture

Sewage is the liquid waste discharged from domestic and industrial sources within an area. It is considered to be a good fertilizer for ponds, having nitrogen and phosphorus as its main constituents. Sewage contains bacteria and protozoa also. Besides carbon and nitrogen traces of zinc, copper, chromium, manganese and nickel are also present, along with some quantity of detergents.

According to Saha *et al.* (1958), sewage of Calcutta has the following composition:

Dissolved oxygen	Nil
Dissolved carbon dioxide	20.96 ppm
H_2S	2.4–4.48 ppm
Alkalinity	170–490 ppm
Free ammonia	12.0–63.6 ppm
Nitrite	0–0.08 ppm

Phosphate	0.01–0.33 ppm
Suspended matter	160–420 ppm
pH	6.9–7.3

Sewage, if used to fertilize ponds without previous treatment, is harmful to fish and other organisms due to low oxygen content, high CO_2, ammonia and H_2S. However, sewage if properly treated is of considerable value in raising the fertility of pond water.

Treatment of Sewage

The sewage is treated by three methods:

1. Mechanical
2. Chemical
3. Biological

The mechanical process consists of screening and filtration, so as to remove coarse suspended matter floating solids including fats and oils, as well as fine suspended material are removed by skimming and sedimentation. Sedimentation is done by letting the sewage into a tank at a high velocity. When the sewage enters a large tank from a sewage channel, there is sudden drop of velocity, resulting in sedimentation.

The chemical method includes coagulation, precipitation, deodorization using chloride, ferric chloride etc, and disinfection by using chlorine, and copper sulphate.

The biological treatment of sewage consists of oxidation of organic matter into CO_2, H_2O, sulphates etc. by using bacteria.

Before sewage is added to a pond it is diluted with freshwater so as to maintain the dissolved oxygen content and reduce CO_2, NH_3, H_3S etc, below the danger level. After sedimentation and dilution, sewage is stored in tanks so as to allow bacterial action for oxidizing organic matter. After sedimentation and dilution at 1: 4 with freshwater, it is added to fishery ponds at monthly intervals. For effective use of sewage and industrial waste, it is necessary to construct, waste stabilization ponds, to retain sewage (10-12 days), till it is made stable and for discharge in to the fishery ponds. As there is considerable fluctuation in the oxygen content in the main stabilization pond, it is stocked with air breathing fishes such as *Clarias*, *Heteropneustes*, *Channa* and *Anabas*.

Major carps are grown in secondary ponds where condition are under control and there is no oxygen deficiency. State Govt. of Calcutta investigations carried out 7.5 cm long fingerlings of rohu, catla and mrigal stocked in the ratio 1:1:1 @ 550 kg/ha gave an annual production of 3,237 kg/ha.

Stocking of fish and the yield in sewage- feed fisheries (saha-1970).

Species	*Ratio of Stocking (per cent)*			
	I	*II*	*III*	*IV*
Rohu	33.3	25	25	12.5
Catla	33.3	50	25	75
Mrigal	33.3	25	50	12.5
Total annual output (kg/ha)	1,850	2,125	1,375	1,525

Best results were obtained when the ratio of stocking of rohu, catla and mrigal was 1:2:1. As a stocking density of 50,000 fingerlings/ha of Indian major carps a gross production of 7,676 kg/ha of marketable fish was obtained in seven months. Ghosh *et al.* (1979) sewage irrigated pond (0.076 ha) for massive tilapia production ranging 70 to 200 mm in length were harvested production 9,350 kg/ha/annum. Thus, use of sewage in fish culture is highly profitable, as the production cost is considerably reduced as fertilizers and artificial food is not required and the yield is higher.

However, there is the possibility of containing human pathogens, accumulation of pesticides and heavy metals. Therefore, Central Institute of Freshwater Aquaculture has designed and developed an Aquaculture Sewage Treatment Plant (ASTP) to include duckweed cultivation along with fish culture. For this a set of duckweed are allowed to grow. These plants help in removing heavy metals and other harmful chemicals and provide oxygen through photosynthesis. Plants used for domestic sewage are *Azolla, Spirodela, Wolffia* and *Lemna*. The harvested weeds can be used for feeding grass carp in the marketing ponds, and excess of weeds can be converted to fertilizer as compost.

After retaining sewage water in duckweed ponds for 2-3 days, it is led to the fish ponds (50 m x 20 m x 2 m). Sewage-fed ponds are

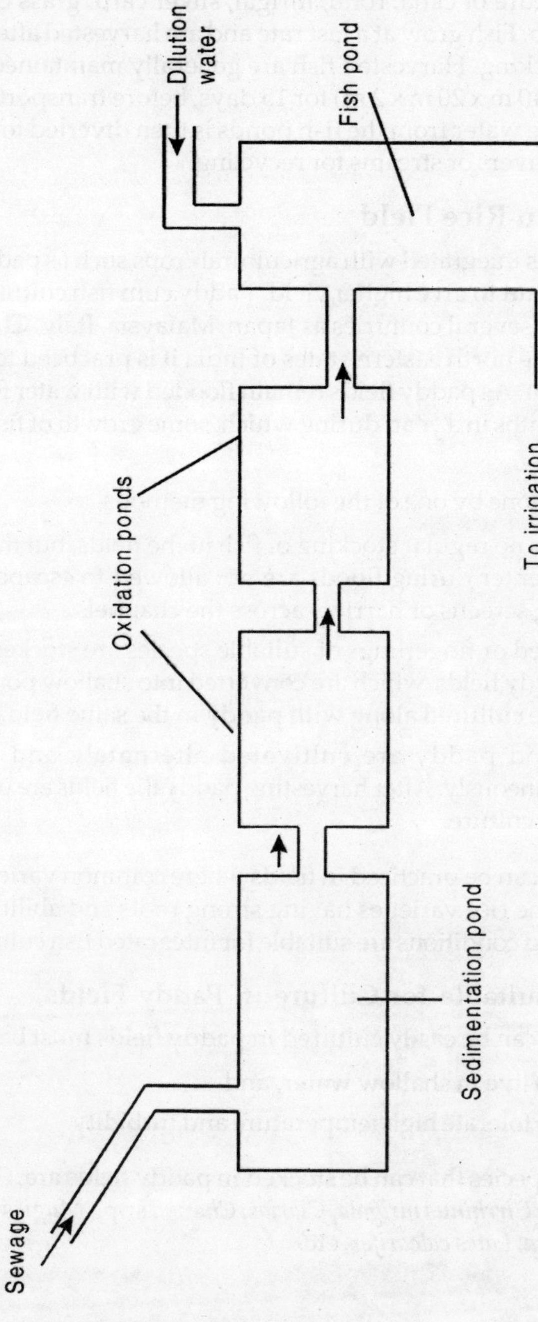

Figure 20.2: Ponds for Use of Sewage in Fish Culture

used for polyculture of catla, rohu, mrigal, silver carp, grass carp and common carp. Fish grow at a fast rate and are harvested after 8-12 months of stocking. Harvested fish are generally maintained in stocking ponds (40 m x 20 m x 2 m) for 15 days, before transporting to the market. The water from the fish ponds is then diverted to the natural water of rivers or streams for recycling.

Fish Culture in Rice Field

Fish culture is integrated with agricultural crops such as paddy, banana and coconut to give higher yield. Paddy cum fish culture is an old practice in several countries as Japan, Malaysia, Italy, China and India. In some north eastern states of India it is practiced to an appreciable extent. As paddy fields remain flooded with water for a period of 3–8 months in a year, during which, some growth of fish is easily possible.

This can be done by one of the following methods.

1. There is no regular stocking of fish in the fields, but those which enter during floods are not allowed to escape by putting screens or barriers across the channels.
2. Fish seed or fingerlings of suitable species are stocked in the paddy fields, which are converted into shallow ponds. Fish are cultured along with paddy in the same field.
3. Fish and paddy are cultivated alternately and not simultaneously. After harvesting paddy the fields are used for fish culture.

Fish culture can be practiced in fields where common varieties are cultivated. The rice varieties having strong roots and ability to withstand flooded conditions are suitable for integrated fish culture.

Fish Species Suitable for Culture in Paddy Fields

Species that can be easily cultured in paddy fields must be:

1. Able to live in shallow water, and
2. Able to tolerate high temperature and turbidity.

Hence, the species that can be stocked in paddy fields are: *Catla catla*, *Labeo rohita*, *Cirrhinus mrigala*, *Clarias*, *Channa* spp., *Mugil* spp., *Tilapia mossambica*, *Lates calcarifer*, etc.

Simultaneous Culture of Fish and Paddy

Various techniques are used for fish culture in paddy fields depending upon the species of fish to be cultivated, varieties of paddy and local climatic conditions.

Four rice plots, 25 x 10 m each may be prepared in an area. Each rice plot is provided with strong barrier wall (dike), 0.3 m high and 0.3 m wide, to prevent leakage of water, and a ditch of 0.75 m width and 0.5 m deep is also provided for fish. The ditches are connected to the main water supply canal through inlet and outlet structures having bamboo screen to prevent entry of wild fish. The water depth in the field varies from 5-25 cm depending on the species of fish and the variety of rice to be cultivated.

Fish fry (1 cm) are stocked @ 5000/ha or fingerlings (8 cm) @ 2000/ha, are stocked five days after transplantation of rice. The natural food of fry and fingerlings in paddy fields consists of plankton, small crustaceans and insects. The quantity of manure normally applied to the paddy field is increased by 50 per cent to enhance production of food for fish use of night soil, compost, stable manure increases fish food and brings about rapid growth. Harvesting of fish is done about a week before the harvest of rice, usually after 2-3 months of stocking. An yield of 500 kg/ha of fish has been reported under simultaneous culture. Fish culture is beneficial to paddy also to some extent fish perform tillage destroy weed and insects that cause damage to the paddy plants, this increasing paddy production. The fish movements help to increase filtering action which cause loosening of soil. The excreta of fish good manure for rice crop. Rice yield increases by 10-15 per cent. Experiments conducted on culture of *Channa striatus* is paddy fields, have shown that 7-13 per cent increase in the yield of paddy, and an average production of 112 kg/ha of fish.

Advantages

1. Fish production is done at no additional cost.
2. Rice yield is increased by 5-15 per cent due to increase in fertilization by fish excreta.
3. Unwanted aquatic weeds and algae are eliminated by fish.
4. Insect pests of rice like the stem borers are eaten by fish. Insect larvae causing malaria are consumed by fish, thus controlling the disease.

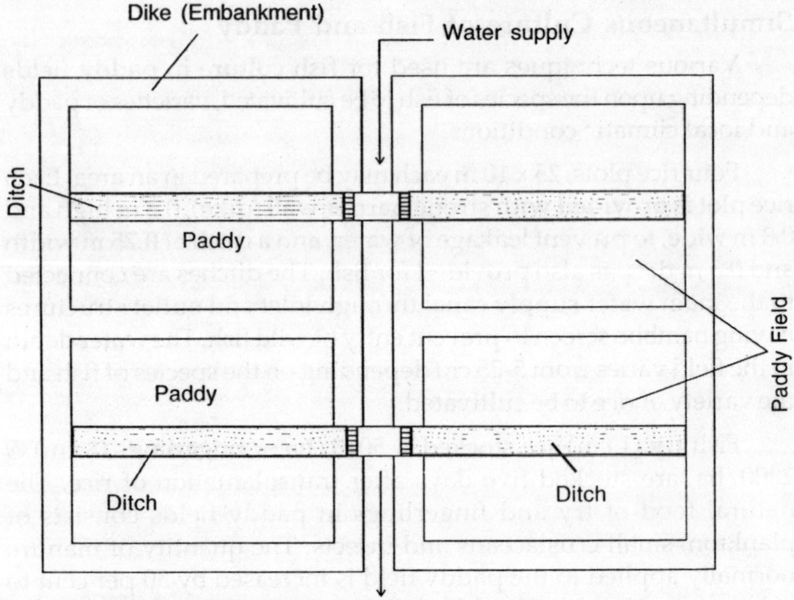

Figure 20.3: Paddy Field with Ditches for Fish Culture

5. Fingerlings, if produced in large number can be sold in the market.
6. Rural agricultural economy is largely strengthened by this practice.
7. Rice and fish are cultivated either simultaneous or alternately in the same water mass.

Rotational Culture of Rice and Fish

In this method fish is cultured after harvesting the paddy, and the field is converted into a temporary pond, one or two weeks after the harvest. Carp fry are introduced @ 20,000/ha, and harvested after 10-12 weeks. Water depth up to 60 cm can be maintained throughout this period, and insecticides and herbicides can be used. An yield up to 2000 kg/ha is possible with proper management.

Chapter 21
Fish Farming with Livestock

A fish culture is combined with livestock for full and maximum utilization of resources. It increases productivity and gives higher income from small land holdings.

Integrated system are ideal for location rural areas easier for farmers to understand and operate. Combined culture systems are energy efficient food production systems and by recycling organic wastes, valuable resource in the form of animal protein can be recovered by its production is enhanced from the unit area.

Fish-cum-Duck Culture

Fish culture in combination with duck rearing is highly profitable, as it enhances the production of both the animals and decreases expenditure on fertilizers for the pond. Duck droppings act as a good fertilizer and are used as food by fish. The duck droppings contain 25 per cent organic and 20 per cent inorganic substances, and are rich in carbon, nitrogen, phosphorus, potassium, calcium, etc. Ducks are called 'living manuring machines' and provide very good fertilizer to the fish pond and dabbling of ducks at the bottom in search of food releases nutrients from the soil, further enhancing production of pond. They feed on insects, larvae, tadpoles, molluscs, worms and aquatic weeds, which do not form food of stocked fish, and thus the two are highly compatible. However, for fish cum duck farming, it is better to stock advanced fingerlings

Figure 21.1: Fish-cum-Duck Culture

(over 10 cm size), as the ducks might feed on smaller ones. Fingerlings are stocked @ 6,000/ha.

A low cost duck house can be constructed on the pond embankment using split bamboo, or a floating duck house can be provided on the water surface. The ducks are allowed free range over the pond from morning to evening, during which they distribute droppings all over the pond. Night droppings are collected from the duck house and applied to the pond every morning. There is no need of additional fertilizer. By stocking 200-300 ducks/ha, 10,000-15,000 kg of droppings are available for the pond in one year.

Good, hardy, breeds of ducks should be selected and the duck house should not be overcrowded, and must be kept clean and dry. 2- 4 months old ducklings are kept on the pond. Natural food may not be enough for 200 ducks, hence artificial feed is to be given every day. Ducks start laying at six months of age and continued for 2 years. Eggs are laid at night and are collected in the morning. Ducks sold after completing 2 years, as the egg laying capacity decreases after 2 years.

Duck droppings cause enhanced production of both phytoplankton and zooplankton, hence catla, common carp and mrigala are suitable species for culture @ 20,000/ha. By integrated fish and duck culture, it is possible to obtain 3500–4000 kg fish, 18000 eggs and 500–600 kg duck meat from one hectare of thus ensuring high profit with less investment. It is highly profitable as it greatly enhances the animal protein production in terms of fish and duck per unit area. Besides manuring, the ducks eradicate the unwanted insects, snails and their larvae which may be vectors of fish pathogenic organisms and waterborne diseases causing organisms infecting human beings.

Fish-cum-Poultry Culture

The droppings of poultry is rich in nitrogen and phosphorus and is used as a fertilizer for fish ponds. Poultry house may be constructed above the water level using bamboo poles, and the droppings may fertilize the pond directly.

The enhancement of meat and egg production in poultry depends upon the breed, variety and strain of the birds, as well as on balanced feeding, health care and good management. The selected birds are kept in pens @ 250 birds/pen, on the floor covered with litter. Dry copped straw, leaves, hay, ground nut shells, broken maize stalks, saw dust, etc is used to cover the floor of the pen up to a depth of 15 cm. The birds kept on this litter and the droppings gradually combine with the material by bacterial action, and from deep litter in about two months, which is very good fertilizer containing 3 per cent nitrogen, 2 per cent phosphate, and 2 per cent potash. About 500 birds produce enough litter to fertilize one ha of water and poly fish culture. Balanced feed is given to the chicks. Eggs are collected every morning and evening, and the birds are sold 18 months of egg laying. Under the integrated fish cum poultry farming, 4,000–4,500 kg fish, more than 70,000 eggs and 1,250 kg chicken meat can be produced from one ha of pond in one year. Chemical fertilizers or supplementary feed for fish are not required. Fully built up deep litter is removed from the poultry pens and stored at a suitable place. It is applied to the pond @ 50 kg/ha/day every morning.

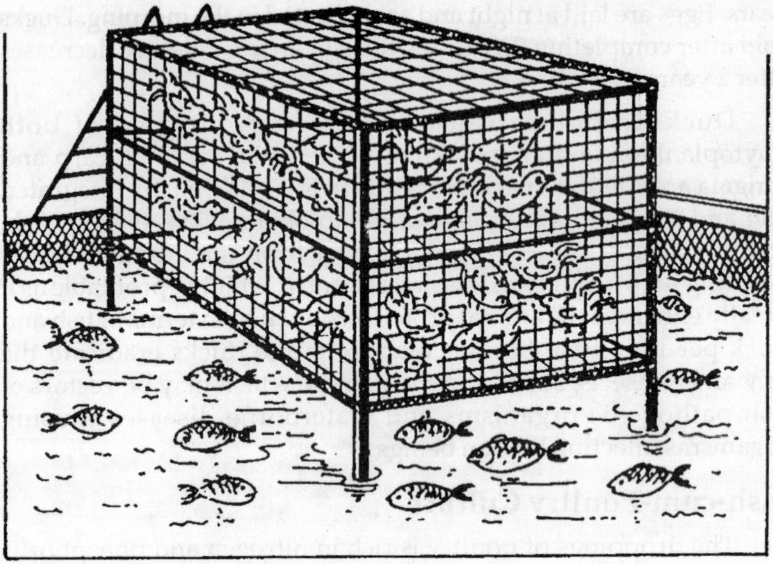

Figure 21.2: Fish-cum-Poultry Culture

It has been estimated that 40 laying birds yield about one ton of manure in an year. Manuring the pond with poultry droppings is found to produce dense population of both phytoplankton and zooplankton dominated by rotifers. Peak plankton production is observed on 12th after manure application. Poultry manure is a complete fertilizer which is characteristic of both organic and chemical fertilizers.

Fish-cum-Pig Culture

Fish culture is linked with pig husbandry by providing pig-houses on the pond embankment or near the pond so that the waste and excreta is directly drained into pond or carried from the animal house to the pond. The pig dung is a good fertilizer. Fish also feed directly on pig excreta, which contains about 70 per cent digestible food for fish. Thus, the expenditure on fertilizer and artificial feed for fish culture is considerably reduced.

Use of pig dung to the pond causes heavy growth of phytoplanktons followed by zooplanktons, and there is considerable increase in detritus at the bottom of the pond. The growth of macrovegetation, which serves as food for grass carp, whose semi digested excreta is used by bottom feeders as food. Hence, polyculture of catla, rohu, mrigal, silver carp, grass carp and common carp, gives high yield ranging from 6,000–7,000 kg/ha/yr, along with 4,200–4,500 kg pig meat. Fingerlings are stocked @ 8,000–8,500/ha.

Pig houses can be constructed in a single row fencing the pond on the embankment using bamboos. These should be well ventilated with cemented floor. An enclosed run is attached to the pig houses, so that the animals get sufficient air, sunlight, exercise and dunging space. The floor is connected to the pond by a drain. Pigs of good exotic variety should be preferred, as they grow fast and the prolific breeders, attaining maturity in six months. About 30–40 pigs are considered enough for one ha of water area, from which two crops of pigs and one crop of fish can be obtained in one year.

The fish feed the excreta of pig and their spilled over food, and the pig dung acts as a fertilizer; hence cost of fish production is reduced. Due to abundance of natural food, the growth of fish is very fast. There is no need of extra space for piggary, hence there is high profit with less investment.

Pig dung contains more than 70 per cent digestible feed for fish. The undigested solids present in the pig dung also serve as direct food source of tilapia and common carp. A density of 60–100 pigs has been found to be enough to fertilize a fish pond of one hectare area. The optimum dose of pig manure per ha has been estimated may be obtained from 50 pigs. If the manure is to be obtained from a dosage of 400 kg/ha/day for 12 times in a year will be required. Fish like grass carp, silver carp and common carp (1:2:1) are suitable for integration with pigs.

Fish-cum-Cattle Culture

Cowsheds may be constructed near the fish pond. Urine and dung is first transferred to oxidation tanks of biogas plant producing methane (CH_4) for household use. The liquid manure from the tank is added to the ponds through small channels.

In an experiment conducted in freshwater tanks in Jammu, 8,000 fingerlings/ha of *Labeo rohita, Cirrhinus mrigala, Catla catla, Hypophthalmicthys molitrix* and *Cyprinus carpio* were reared. Cattle manure @ 7,000 kg/ha was used in installments. Supplementary food consisting of rice bran and mustard oil cake (1:1) was given @ 3 per cent of the total body weight in one of the experimental tanks. A production of 4779 kg/ha/yr was obtained in the tank with supplementary feed, and 2,334 kg/ha/yr in the pond without supplementary food.

Fish-cum-Poultry-cum-Pig Culture

All the three animals can be integrated for better production. Chick droppings are used as food by pigs, hence the two are reared in a two-tier house, pigs occupying the lower chambers and the chick housed in upper part. Such house can be constructed on the bank of the pond at a nearby place.

Thus, fish culture integrated with pig, duck, poultry or cattle provides a higher source of income to the farmer having a small land holding. These practices help in realizing the full potential of aquaculture development in the country. It is expected that integrated fish farming practices will increase in future in suitable agroclimatic regions of the country as they are dependent on eco-friendly measures and ensure higher returns as well as sustained production levels of fish and other animals.

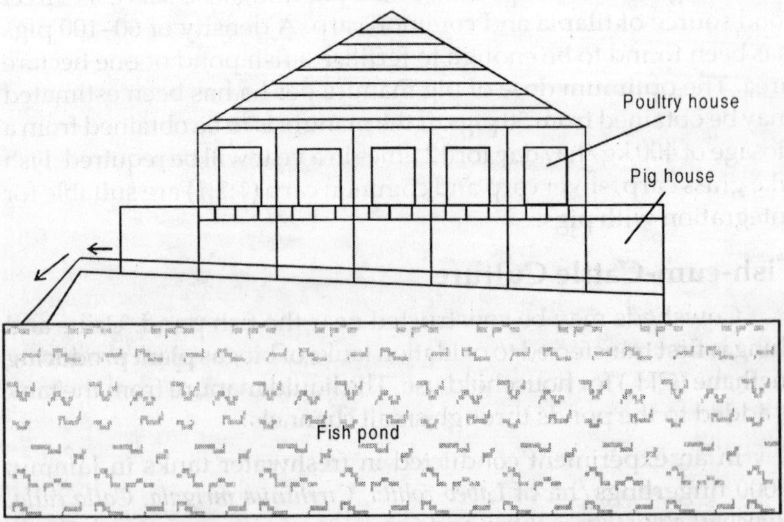

Figure 21.3: Fish-cum-Poultry-cum-Pig Culture

Chapter 22
Brackish Water Fish Culture

Brackish water is formed by mixing of fresh and sea water near the along the coast where river is meet to the sea. The characteristic brackish water has high salinity, rich in nutrients, the *p*H ranges from 7.5–8.3. The water contains several species of algae and other aquatic plants. Zooplankton and larval stages of crustaceans and fish are dominant. Water is well oxygenated in shallow areas.

In the country an area about 2 million ha is available for brackish water farming. The fish farming in brackish waters is a deep-rooted and well organized industry in southeast Asia. Vast area with rich resources suitable for brackish water fish farming is available in our country. The coastal area of Bay of Bengal and the Arabean Sea are very much suitable for development of brackish water aquaculture.

Fishes of the brackish water are mainly marine, which can tolerate large salinity changes and are called euryhaline. The milk fish *Chanos chanos* and the mullets (*Mugil* spp.) are the two principal fishes used for cultivation either as monoculture or as polyculture when milk fish is raised with *Tilapia* or *Puntius*. Both milk fish and mullets can tolerate wide range of salinity and temperature. Growth is rapid. However, these breed in sea only, and therefore, the seed has to be collected from there for introduction into the pond.

The *bheri fish culture* in Sunderbans of West Bengal, the prawn culture in *Pokkali* paddy fields of Kerala and the *culture in coastal lagoons* are the most important types of brackish water fish culture practiced in India.

The great estuaries on river mouths and backwaters offer important potential of fish culture of particulate species. The fishes generally raised in brackish waters include the indigenous milk fish (*Chanos chanos*), pearl spot (*Eutroplus suratensis*), bhekti (*Lates calcarifer*), tarpon (*Megalops cyprinoids*), and the exotic tilapia (*Tilapia mossambica*) and gouramy (*Osphronemus gouramy*). The practice is traditional. Special mention may be made of three cases:

1. The *Bhasabandha* or *Bheri* culture operates in Sunderbans and other parts of West Bengal. The *Bheries* are low lying fields enclosed by erecting bundhs (embankments) or utilizing natural embankments. Slucie gates are installed for control. High tides and inflowing rivers bring water and fish seed of great variety into the bheris. The embankments guards against the ill effects of flood and tides. Split-bamboo screens are used at the sluice gates to prevent escape of fish, deliberate or forced under receding tide. Box-shaped traps are attached to the screens to collect larger specimens. The operation begins in January-February when the bheries are filled with water and fish seed. The fishes grow to marketable size by September, when harvesting begins. Little management is undertaken and, therefore, investment is very low. However, the yield is also very low about 100-200 kg/ha/year. The fishes composing the catch include *Lates calcarifer, Mugil spp., Mystus gulio, Polynemus tetradactylus, Scatophagus argus, Anguilla bengalensis* and *Glossogobius* spp.
2. The *Narakkal Swamp waste* in Kerala functions as a highly developed mulletry. Mullet fry are allowed to enter the

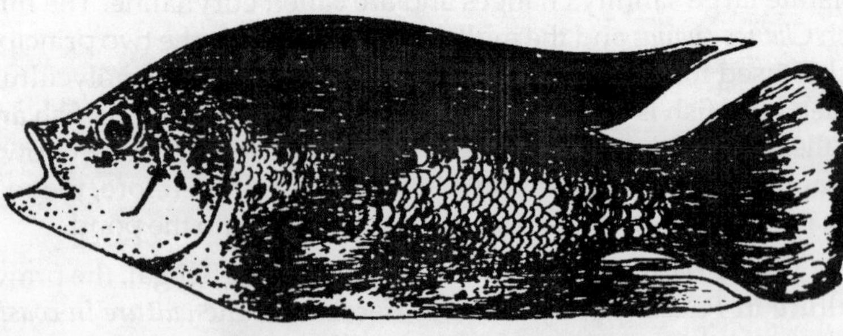

Figure 22.1: ***Tilapia mossambica***

fields with tidal water. These are then raised in enclosures bound by embankments and sluice gates.

3. The Kerala backwaters (Pokkali fields) are the other important sources. Besides the dominant prawn farming, rearing of mullets, pearl spots and tilapias is also carried out.

Fish Culture

The seed of euryhaline species are collected from estuarine sources, and are acclimatized before stocking. In Bengal, fish are cultured in embanked brackish water areas, known as *'Bhasa-bhasa'* fisheries or Bheries. River beds and creeks in the estuarine areas get silted up due to the action of tides, and in due course such areas are reclaimed for agricultural purposes by constructing bundhs or embankments to keep away flood and tidal water. Some portions of such reclaimed areas are too low for cultivation, and are usually utilized for fish farming. Many of the reclaimed areas are provided with an additional outer bundh, and the low lying land in between in used for fish culture. Water is let in during high tide through impoverished or sluice gates and brings in several species of fishes and fish seed. In the channel inside the sluice gate, split bamboo screens, shaped like an inverted V or W, are fixed in such a way that they allow admission of fish during high tide, but do not allow escape of fish when the tide recedes. Box-like traps are constructed with the V-shaped screens and larger fish are collected in them, of which the useful varieties are transferred to stocking ponds.

Fish culture in bheries begins in January and February, when ponds are stocked with fry. After stocking no attention is paid by the fisherman to the bheries. Fishing is generally done in winter, when the fishes have grown to marketable size. The main species obtained from bheries are *Lates calcarifer*, *Mugil* spp., etc.

Bheries simply need construction of proper embankment and sluice gates. Selective stocking and rearing of fry is not practiced, hence the yield is generally low. According to Pillay (1954), if the predatory and weed fishes are not allowed to enter into the bheries, and proper fertilization is done, the yield can be increased considerably.

Brackish Water Farms

There are only a few farms along the coast of Kerala. These are situated at Narakkal, Malippuram, Kadappuram and Eranholi. The Narakkal, fish farm was originally a swampy waste but have now been converted into a fine mullet farm and consists of several ponds with earthen bundhs and wooden sluice gates. The fry are let in along with the tide during the monsoon. Selective stocking is also practiced in these farms and the seed is collected from channels and creeks adjoining the farm. During summer, the water level goes down considerably resulting in an increase in salinity and water temperature. During this period, the fishes take shelter in deep

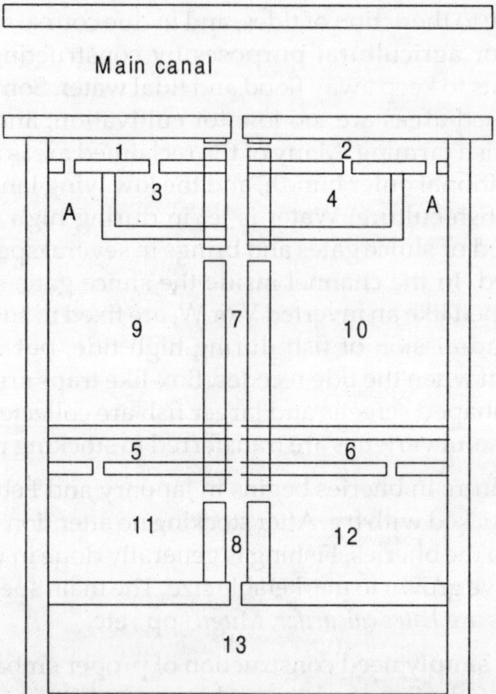

Figure 22.2: An Ideal Brackish Water Fish Farm 1&2: Wintering ponds; 3&4: Nursery ponds; 5&6: Passages ways; 7: Subcanal; 8: Subcanal & passage ways; 9 to 13: Production ponds; A: Acclimatization pond

ditches dug for the purpose. Species cultured in these farms are *Mugil* spp., *Etroplus*, *Chanos*, *Polynemus*, and *Tilapia*. *Mugil cephalus* constitutes 75 per cent of the stock and grows to 1 ½ kg in a year.

Historically, brackish water fish culture appears to have developed in India and some other countries. In India little attention has been paid in developing coastal farms on modern lines. The brackish water farm at Kakdwip near Calcutta is one of the modern farms with nursery, rearing and stocking ponds.

For development of fish farms along the coastal lines, it is necessary to examine the soil type, water supply and the available fish seed in the area.

Brackish water farm at Kakdwip near Calcutta has been developed along modern lines and has nursery, rearing and stocking ponds. Fish fry enter the farms during high tide. Selective stocking is done and fish seed are collected from channels and creeks.

Ecology of Brackish Water Ponds

Water

As a medium for fish culture, brackish water has many advantages. It contains a rich supply of nutrient salts; it is perfectly buffered against abrupt changes in *p*H.

Schuster (1952) considers a tidal range of 1.5 m to be essential for a brackish water fish farm. Water in brackish water ponds is always alkaline and there is very little variation in *p*H. Some of the more important physico-chemical characteristics of water in a modern brackishwater farm at Kakdwip in Lower Sunderbans, West Bengal, observed during different months of the year, show fluctuation in salinity between 2.19 per cent and 22.41 per cent, that of *p*H between 8.0 and 9.2 and dissolved oxygen between 6 ppm and 14.0 ppm.

Soil

The quality of soil underlying an estuarine pond is to some extent dependent on the nature of the catchments area of the river. Physical and chemical characteristics of the farm soil of Kakdwip, Sunderbans, West Bengal, as given below:

Sand	8 per cent
Silt	35 per cent

Clay	55 per cent
Organic carbon	0.34 per cent
Salinity	1.0 per cent
Free $CaCO_3$	3.46 per cent
Carbonate	0.05 meq per cent
Bicarbonate	2.00 meq per cent
Total nitrogen	0.069 per cent
Available nitrogen	18.5 mg/100 g
Available phosphorus	4.0 mg/100 g

Selection of Species

The principles of selection of fish for culture in brackish water ponds are the same as those for freshwater ponds except that the species should be euryhaline, able to adjust themselves in an environment of fluctuating salinity. The basic consideration in the selection of culturable species are as follows:

1. Rapid growth potential.
2. Ability to use the natural food of the pond efficiently and accept artificial feed with low conversion ratio.
3. Small head, high body and thick back.
4. Hardy and resistant to disease.
5. Prolific and easy to breed in ponds.
6. A prolonged breeding season or multiple breeding frequency.
7. It should be non-predaceous, plankophagous and preferably herbivorous.
8. It should be companionable and compatible with other fishes which may be cultured together
9. It should have low bone to flesh ratio, and
10. Palatable with high nutritive value.

The important cultivated Indian brackishwater fishes are: *Elops* spp., *Chanos chanos*, *Mugil* spp., *Lates calcarifer*, *Etroplus suratensis*, *Tilapia mossambica*, etc.

Mullets

Mullets are the most important of all the culturable brackish water species. Several species of mullets are found in the brackish water. These are *Mugil cephalus, M. tade, M. speigleri, M. passia. Mugil cephalus* forms a large part of the fishery of estuarine water.

Feed

Fry: Phytoplankton and zooplankton with diatoms and epiphytic *Cyanophyceae* as main component of diet.

Fingerlings: Detritus and diatoms.

Adults: Diatoms, blue-green algae, green algae, detritus, small crustaceans etc. When stocked in ponds, consumes considerable quantities of decayed plants.

Milk Fish

Chanos chanos is an important species of the brackish water.

Feed: *In sea*: Benthic microscopic vegetable organisms, mollusks and dead decapods.

In estuaries: fry and fingerlings feed on *Bacillariophyceae*, and broken down chlorophyceae. Adults feed mainly on diatoms, algae, lamellibranches, fish eggs, etc.

In Brackishwater ponds: Fry and fingerlings feed on *lab-lab*. Juveniles (50-100 mm) feed on blue green algae, diatoms, protozoans and microcrustaceans mixed with mineral particles and organic detritus. Adults subsist on benthic food.

Milk-fish fry are also acclimatized to water of varying salinity for about 6 hours.

Fry of milk fish, *Chanos chanos* is obtained extensively on the East coast of India from Visakhapatnam southwards through East Godavari, Krishna, Guntur and Nellore in Andhra Pradesh, etc. The most important center in Tamil Nadu for collecting milk fish fry is at Pamban (Ramnad district). Fry caught usually during April to June.

Milk fish fry have a tendency to swarm in large shoals in shallow tidal creeks having calm and clear water. The gear consists of a piece of cloth measuring 3 m x 1.2 m and 3 m long rope to serve as a scare-line, with palmyra leaves attached to it at short intervals. The scare-line is dragged in a semicircular pattern and is closely followed by two fisherman holding a stretched strip of cloth to catch the enmeshed fingerlings escaping from the scare-line.

Procurement of Stocking Material of Brackishwater Fishes

Mullets are probably the most important of all the cultuarable brackishwater fishes. The fry of mullets migrate into the brackish water areas, and are caught for culture along the tidal creeks and pools. Their fry are collected in shallow coastal waters, estuaries, tidal channels, burrow pits fed by tidal canals and lagoons. The fry and fingerlings of mullets are caught by hapa nets, which are rectangular pieces of mosquito net cloth. Small triangular or circular dip nets are also used. Fry of *M. cephalus* and *M. macrolepis* are available near the mouth of Chilka lake. These fry are used for culture in brackishwater ponds. Fry of the *Mugil* spp., mostly 1.8 to 5.0 cm long, are trapped along the Contai caost (West Bengal) in tidal creeks and pools. In the Sunderbans area, they are collected in burrow pit which retain the fry as the tide recedes. Main centers of collection are: Muriganga and Baratala rivers at Kakdwip, Kachuberia near Sagore, Frazerganj, Bokkhali, Dhablot near Sagor, Monkhali at Canning, Namkhana, Raidighi and Amlamethi on Bidya river. Fry of *M. parsia* also occurs in the Mahanadi Estuary during the periods January to May and October to December and in the Maharashtra coast during January and June to September.

Mullet fry are caught by *hapa* net, a simple rectangular cloth piece or a rectangular bag (mouth 1 m x 0.5 m and length 1 m) made of mosquito netting. Fyke nets are also used to catch mullet fry. In Tamil Nadu, mullet fry and fingerlings are collected with fine meshed drag nets. The fry of mullets are exposed to water of decreasing salinity for a period varying from 1-5 days. Mullets are generally caught in gill nets and seine nets, and in shallow areas by cast net.

Induced Breeding of Mullets and Milk Fish

Most of the fishes cultivated in brackishwater are known to breed naturally in the sea. The seeds that enter the estuaries, coastal lakes and backwaters, use these water-bodies as nurseries for feeding and growth and migrate back to the sea on attaining certain sizes and certain stage of maturity for purpose of breeding.

Among the cultivated mullets, *Mugil cephalus, M. tade, M. parsia* and *M. macrolepis* attain maturity under culture conditions. Expériments on induced breeding of mullets were initiated in India during 1961 and success was achieved in breeding *M. cephalus* by hypophysation (Barrackpore, 1962). Successful breeding and rearing of larvae of *Mugil macrolepis* have been reported by Alikunhi *et al.* (1971) and Sebastian and Nair (1975). Chaudhuri *et al.* (1977) conducted experiments on induced breeding of *M. cephalus* by pituitary hormone injections and obtained 26,050 hatchlings from 8 spawners. The dose gonadotropin required varies between 12 and 21 mg/g body weight of the female. The total dose is applied in two injections: one-third of the total dose is followed 48 hours latter by the remaining two-thirds. Abdominal distension is first observed 8 hours after the second injection. Two or three mature males are placed in a 50 gallon aquarium with a female about 10 hours after the second injection. Males are not given injection. The first release of a small number of ripe eggs stimulates the males to liberate spermatozoa and fertilization follows.

Spawning of milk fish, *Chanos chanos*, the only successful breeding so far obtained is from Philippines. A breakthrough achieved in milkfish breeding through induced spawning in the Philippines (Anon., 1977). Mature specimens, but males and females, were obtained from the sea. The hormones employed were 4 ml saline solution containing 60 mg acetone dried pacific salmon pituitaries and 4000 IU of HCG in the first injection. The fish were anaesthetized with quinaldine sulphate, or 2-phenoexythanol, after which the second injection comprising 6 ml hormone solution was administered. The female specimen was killed and the eggs were removed to be fertilized by dry and wet methods with sperms obtained from mature males. Dry method of fecundation has been found more effective in milkfish.

At 31 per cent $_0$ salinity and water temperature in the range of 26.5 to 28.9°C, the hatching of eggs reported to occur 35 hours after fertilization. So far natural spawning of milkfish has not been reported in induced breeding by hypophysation.

Hatchery Management

Mullets

Breeding is relatively easy, hatching and rearing of larvae in commercial quantities pose problems. Fertilized eggs are to be kept in round plastic tanks of 0.5 to 1.0 ton capacity, with either flowing or stagnant sea water. In the flowing water type, two to three conical hatching nets are hung in the water and their tail ends connected with plastic pipes to supply a continuous upwards flow of fresh sea water so that the eggs would continue to roll and even buoy up within the net. The stagnant water tanks are to be kept well aerated and the water changed whenever necessary. Hatching of eggs observed in 32 to 52 hrs at temperatures 19.5°C to 28°C.

During early development, the embryonic yolk sac and oil globules are utilized for larval nutrition but after the third day, the mouths of the larvae open and extraneous feeding has to start. For feeding the larvae, oyster Veligers, trochophore larvae, rotifers, copepod and *Artemia* nauplii were used. The important guidelines by Liao *et al*. (1972) for rearing mullet larvae are: i) larvae should be hatched in tanks with aerated water prepared for rearing; ii) they should not be transferred right after hatching; iii) seasoned and stable sea water should be used for rearing which should be gradually freshened; iv) sharp changes in water temperature and its quality should be avoided; and v) the nature of the food added should match with the stages of development of the larvae.

Modern Brackishwater Fish Farm in India

The first farm was constructed near Tuticorin followed by another one on Krusadai Islands. The important brackishwater fish farms in the State of Kerala are at Narakkal, Ayiramthengu, Malippurum, Kadappuram and Eranholi.

The Narakkal fish farm site, originally a swampy waste, has been converted into a mulletry consisting of several ponds with earthen bundhs and provided with wooden sluice gates having adjustable planks and split bamboo gratings for regulating the flow of water (Pillay, 1948). Apart from the conventional method of intake of fish seed with the tidal flow, supplementary selective stocking is also practiced in these farms, the source of stocking material being channels and creeks adjoining the farms. The farm ponds are neither manured nor is artificial feed given. The annual variation in salinity is in the range 2 per cent $_0$ to 32 per cent $_0$ and the water temperatures as high as 31°C.

In summer, when the water level drops down considerably with consequent increase in salinity and temperature, the fish take refuge in ditches especially provided in the pond substrate. The areas of Narakkal, Ayirmthengu and Malippuram fish farms are 18.40 ha, 20 ha and 21 ha, respectively. Brackishwater species that thrive best in these farms are milkfish, mullet fish, pearl spot, and *Tilapia mossambica*.

The brackishwater farm at Kakdwip (Calcutta), is located on a 14 ha plot of land and has arrangements for automatic influx of fish and shrimp seed along with the tidal water. However, the seed are not allowed to enter the ponds as the water is filtered through suitable screens placed at individual inlets to ponds. Stocking is done with segregated seed of selected species at predetermine densities. The farm was originally designed with 40 nursery ponds of 22 m x 6 m, two rearing ponds of 51 m x 19 m and one stocking pond of 43 m x 22 m. Culture operations are done in two stages. In the first, nursery–rearing stage, the seed collected from natural sources are reared for 1-2 months in specially prepared nursery ponds before transporting into the second stage production ponds.

In the culture of *Lates calcarifer*, gross production at the rate of 2759.5 kg/ha in 240 days (Jhingran, 1977). Other experimental monoculture operations have yielded production at the rates of 1775 kg/ha/270 days with the mullet, *Liza tade*; 800 kg/ha/140 days, with *L. parsia* and 710 kg/ha/150 days with milk fish, *Chanos chanos*. Mixed culture of mullets, *L. tade* and *L. parsia* has resulted in production at the rate of 2200 kg/ha/yr. A successful combination of mullets, milkfish and prawns under polyculture is reported to yield up to 2579.8 kg of fish per ha per 9 months (Jhingran, 1977).

Monosex Culture

Sexes differ in their rate of growth, hence commercial fish production of the sex having better food conversion efficiency would be profitable. Production of only male or female fish can be done by one of the following methods:

1. Male and female fish are identified before maturity and segregated in tanks. The faster growing sex is reared to

full size for marketing. For example in Tilapia, there is a clear sexual dimorphism rendering the sexing and sorting easily possible. The males have a single urinogenital opening on the tip of a papilla while the female has separate openings.

2. By using sex hormones: A reliable method is to feed fry with minute dose of sex hormones so as to produce all males or females. The gonad of a teleost is undifferentiated in the beginning, and can be triggered to give rise to ovary or testis by treating with hormones. However, success depends on the use of correct dose at the proper time, and for the required period.
3. By causing sex reversal: Sex reversal has been successful in several species as *tilapia*, *oryzias* and *Carassius auratus*. In *tilapia*, male grows faster than the female. Administration of 15-16 mg methyl-testosterone/kg food has produced all males in this species. In Israel, newly hatched fry were given methyl-testosterone in food for 5 weeks which resulted in the production of males. In salmon and trout, all females were produced by administration of *17-b* oestradiol @ 20 mg/kg food, for a period up to 60 days. In other species of fish, similar methods can be used to produce only one sex.
4. Another method successfully used is to convert genetically female fish to functional males by feeding methyl testosterone @ 3 mg/kg food. These males are used to fertilize normal eggs and only females are produced as Y-chromosome is absent in the sperms.
5. Experimental hybridization is a relatively easier and more effective way for monosex culture. In tilapia, interspecific and intraspecific mating yields monosex male stocks as follows:

   Male *T. macrohir* x Female *T. nilotica*

   Male *T. hornorum* x Female *T. mossambica*

   Male *T. mossambica* x Female *T. nilotica*

   Male *T. mossambica* x Female *T. mossambica*
   (African) (Malaysian)

Chapter 23

Harvesting and Marketing of Fishes

Harvesting

The labor and equipments are required for harvesting a perennial pond. Large sized full grown fish (25-35 cm or more) are harvested and transported to the market. The ponds are harvested at intervals and larger fish of marketable size are removed and transported to the market. Generally, large sized fishes are caught and smaller ones allowed to grow to provide maximum yield. Sometimes the entire crop is harvested regardless of size. In well organized fisheries, fish below a particular size are not generally captured.

Netting should be done with fine-meshed nets (preferably below 1-2 cm knot to knot), as a bigger mesh-size entangles the small-sized breeders, and the threads of the net may cut through the fins of a breeder and tear off the scales, or a loop of the net may strangle the fish around the neck, *i.e.*, causing gilling. Netting should be done very slowly, so that fishes do not get much chance to jump. After encircling the fishes, the net should be kept double folded in the shallow, marginal water of the pond.

The harvesting should be done when water is 0.3 m deep. The fish should be caught drag net with repeated dragging operations. The live fish should be kept in hapa suspended in water for marketing purpose. Food fish may be harvested daily by netting with gill net (3–4 cm bar gill net). The harvesting fish by drag net

Figure 23.1: Harvesting of Fish

sometimes becomes difficult because of weeds and therefore a long gill net can be used for partial harvest. A record should be kept of the number and weight of fish caught so that production can be assessed and additional stocking programme formulated.

Fish Marketing

The marketing are additional tasks in which the farmer may not have any experience. Therefore, arrangement may be made with a contractor and profits can be divided according to royalty rates mutually agreed upon. If local marketing is possible and profitable, a party of local fisherman may be engaged. A farmer can market the freshly caught fish himself.

This is a very important part of fishery trade the commercial side of it. The success of the trade much depends upon how efficient is the fish marketing system. Simply put, fish marketing finds a good market to sell fish and its byproducts, serving as a direct link between the trader and the consumer. Fish marketing has important role in both the trader and consumer. It must induce the trader to endeavor for maximum production, best of commodity and its timely supply at reasonable cost. This can be achieved only when fish marketing is able to offer a really encouraging market in and thus promotion of the economy of the country need not be emphasized as also its effectiveness in solving to some extent the food problems of countries faced with food shortage.

Some important functions of the fish marketing are listed below:

1. To know when, where and what of the demand position of the consumer market.
2. To assess the demand of any one place as to its amount, kind and time with a view to supply or withhold fish.
3. To regulate supply in such as way as to enable stabilization of prices in periods of bumper landing or extremely poor landing of fish. The distribution system at all levels has to be supervised efficiently to prevent occurrence of local artificial scarcity and consequent dangerous price fall.
4. To determine the proportions of fish landed for diversion to various agencies dealing with preservation, storage, curing yards, canning industries and manufacturing units of various products and by-products, to markets for fresh consumption etc.
5. To establish standards for quality control at various levels of storage, preservation and production of products or by-products in accordance with requirements of local and foreign markets.
6. To prevent losses due to wastage.
7. To keep a strict vigil on unscrupulous elements and their activities in the trade. Middlemen, cooperative societies, money lenders, hoarders and retailers engaged in the trade to be controlled for their harmful activities or functioning.
8. To educate the people about the importance of food value of fish and to popularize, through publicity, in them the habit of eating fish.
9. To explore possibilities of export and import thereby to help the country earn more foreign exchange and cooperate in international trade.

Part II
AQUARIUM KEEPING

Chapter 24
Aquarium Keeping

Introduction

Taking a few moments off daily chore, it is not bad watching a small aquatic home with glassy walls. Colorful enchanting members that enhance the décor of the place are seen moving inside it. These lovable members are, of course, the ornamental fishes inside an aquarium, which represent a natural ecosystem. Watching fishes inside an aquarium refreshes one's mind, remove stress of the modern day life and allows one to deep into a fish's world.

Aquarium keeping is an age old practice and has become one of the most popular hobbies. As on high demand, throughout the world, rearing and marketing of the ornamental fishes for the aquarium can be taken up as a means of amiable livelihood. It was during the Sung dynasty, about 1000 years ago, when the Chinese first began to domesticate gold fishes. Later on these fishes were transported to the other places in the world. Glass aquaria were first used by the Romans. During 18th century, gold fishes were transported to England and keeping them in glass bowls became a popular hobby.

Keeping fish in an aquarium at home is a popular hobby for many people. Usually fish aquarium is kept in the room for decorative purpose, but fish are maintained in the laboratory for scientific studies also. Culture of ornamental fishes also provides jobs to many people

Aquarium Fishes and Plants

Aquarium Fishes

A few fishes of small size are introduced in to the aquarium. Most of the ornamental fishes are from Asian countries. Some of them like Indian glassfish, sand loach, banded loach, dwarf gourami, barbs and orange chromide etc., are of Indian origin. Besides, there are some rare aquarium fishes of Indian origin which have high demand in the world market. It is but always better to begin with inexpensive and hardy aquarium fishes. Moreover, fishes which occupy and feed at different levels of aquarium are the most suitable for nurturing together. Aquarium fishes are broadly divided into two categories: egg layers and live bearers.

Egg Layers

The egg layers aquarium fishes they release the eggs in water. Fertilization is external outside the body. Examples, goldfish, Indian glassfish, Angelfish, Siamese fighting fish, Golden gorami etc.

1. Goldfish (*Carassius auratus*)

Classification

Kingdom: Animalia

Phylum: Chordata

Class: Actinopterygii

Order: Cypriniformes

Family: Cyprinidae

Genus: Carassius

Species: *auratus*

Linnaeus, in 1758, discovered this species. The Goldfish (*Carassius auratus*) one of the most commonly kept aquarium fish. A relatively small member of the carp family (which also includes the koi carp and the crucian carp), the goldfish is a domesticated version of a dark-grey/olive/brown carp native to east Asia (first domesticated in China) that was introduced to Europe in the late 17th century.

Biology

Maximum length of 23 inches (59 cm) and a maximum weight of 9.9 pounds (4.5 kg). In optimal conditions, goldfish may live more than 20 years, most household goldfish generally only live six to eight years.

Aquaria

The goldfish is usually classified as a coldwater fish, and it can live in an unheated aquarium. One square foot of water surface area for every inch of goldfish length (370 cm^2/cm). If the water is being further aerated by way of water pump, filter or fountain, more goldfish may be kept in the container.

Goldfish may be coldwater fish, conversely, temperatures over 25°C (77 °F) can be extremely damaging for goldfish. Common and comet goldfish can survive, and even thrive, in any climate in which a pond for them can be created.

Feeding

Like most fish, goldfish are opportunistic feeders. Goldfish need only be fed as much food as they can consume in three to four minutes, and no more than twice a day. Extreme overfeeding can be fatal, typically by bursting of the intestines. Special goldfish food has a lower protein and higher carbohydrate content in the two form–flakes that float at the top of the aquarium, and pellets that sink slowly to the bottom. Young goldfish also benefit from the addition of brine shrimp to their diet.

Environment

Their native climate is subtropical to tropical and they live in freshwater with a pH of 6.0–8.0, a water hardness of 5.0–19.0 dGH, and a temperature range of 40 to 106 °F (4 to 41 °C). They are omnivorous and do best with a wide variety of fresh vegetables and fruit to supplement a flake or pellet diet staple. When transferring a store-bought goldfish to a pond or a tank, the temperature in the storage container should be equalized by leaving it in the destination container for at least 20 minutes before releasing the goldfish.

Breeding

Goldfish, like all cyprinids, lay eggs. They produce adhesive eggs that attach to aquatic vegetation. The eggs hatch within 48 to

72 hours. Within a week or so, the fry begin to look more like a goldfish in shape. Breeding usually in spring. Eggs should then be separated into another tank, as the parents will likely eat any of their young that they happen upon. Dense plants such as *Cabomba* or *Elodea* or a spawning mop are used to catch the eggs. Males chase the females around, bumping and nudging them in order to prompt the females to release her eggs, which the males then fertilize.

Goldfish and other carp are frequently added to stagnant bodies of water in order to reduce the mosquito populations.

2. Pearl Gourami (*Trichogaster leerii*)

Classification

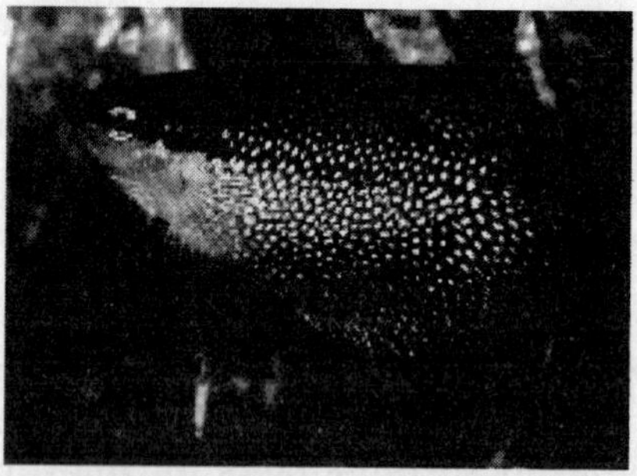

Kingdom: Animalia

Phylum: Chordata

Class: Actinopterygii

Order: Perciformes

Family: Osphronemidae

Genus: Trichogaster

Species: *leerii*

Biology

An elongated, oval shaped fish with lateral compression. The ventral fins are filamentous and the anal fin is runs from just beyond the ventral fins nearly to the fan-shaped caudal fin. The dorsal fin is shorter than that of the previously mentioned *Colisa* species and runs from the mid-back to caudal peduncle. The back of a male is olive brown, while the flanks are brown to brownish red. The flanks are littered with numerous, over-laying dots that are pearl to iridescent violet in color. A black lateral stripe extends from the lips to the caudal peduncle, although in some specimen, this stripe is inconspicuous. The throat and belly are bright orange red to red in color. The fins have the same pattern as the flanks, save for the ventral fins which are orange-red in color. Females are blander, although they also have the pearl pattern over-laying their flanks.

Size

5"(13 cm).

Tank

A 24" (61 cm) or 10-20 gallon (38-76 L) tank is adequate for a single pair. At least a 20 gallon (76 L) is recommended for more. Arrange the tank in dark colors with a cover of floating plants. The tank should be well-planted along the edges, and an open swimming area should be left in the center.

Water

pH 5.5–8, temperature 75-86°F (24-30°C).

Food

Live *tubifex*, insects, insect larvae, crustaceans, flakes, pellets, chopped spinach and lettuce.

Sex

The males are more colorful and have pointed dorsal fins.

Breeding

The water level should be reduced 4-6" (10-15 cm) and no water current should be present. Condition the pair with insect larvae, *Artemia*, and *Daphnia*. The male builds a large bubble nest at the water's surface in floating plants. After an active courtship, 200-300 clear eggs are laid. Spawning may continue for extended periods-even weeks. These float up to the bubble nest, and are corralled by the male. They are guarded by the male. At this time, all other fish should be removed, including the female. The eggs hatch after 20-30 hours. The fry remain in the nest for 4-5 days. Start feeding with paramecia, rotifers, and infusoria. The male can be removed after the fry are free-swimming. The fry are slow-growing. The Pearl Gourami regularly lives over 8 years. A hardy fish suitable for a beginner.

3. Golden Gourami (*Trichogaster trichopterus)*

Classification

Kingdom: Animalia

Phylum: Chordata

Class: Actinopterygii

Order: Perciformes

Family: Osphronemidae

Genus: Trichogaster

Species: *trichopterus*

The three spot gourami (*Trichogaster trichopterus*) originating in southeast Asia. It is a popular freshwater aquarium fish.

Biology

Tank

The ideal tank set-up would be an aquarium of a minimum of 20 gallons and have plenty of live plants as well as rocks and driftwood for use as hiding places.

Diet

The three spot gourami is an omnivore and requires both algae-based foods as well as meaty foods. An algae-based flake food, along with freeze-dried bloodworms, tubifex worms, and brine shrimp will provide these fish with the proper nutrition. The average weight of a blue gourami is 2-3.5 ounces.

Breeding

In the male, the dorsal fin is long and pointed, while the female's is shorter and rounded. When ready to breed, the male builds a bubblenest and then begins to entice the female by swimming back and forth, flaring his fins and raising his tail. When this behavior is noticed, the water level should be reduced to 6 inches. After spawning the female should be removed to a separate aquarium as the male may become aggressive toward her. The male will tend to the eggs until they hatch, and after hatching, there should be frequent water changes, especially during the third week, as this is when the labyrinth organ is developing ("labyrinth organ" (modified gill) which is used like a lung to breath atmospheric air). The fry should be fed infusoria and nauplii.

4. Neon Tetra (*Paracheirodon innesi*)

Classification

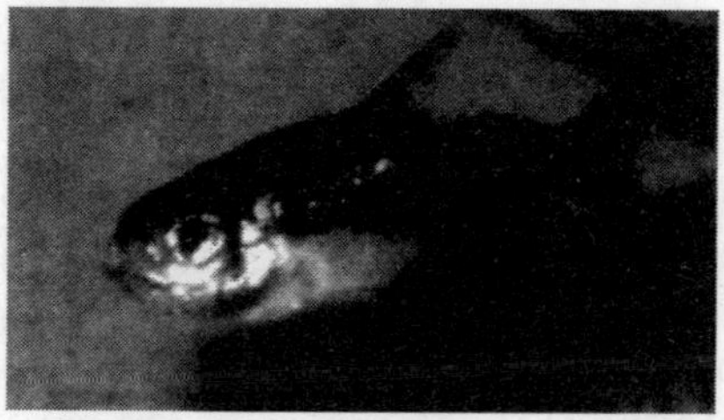

Kingdom: Animalia

Phylum: Chordata

Class: Actinopterygii

Order: Characiformes

Family: Characidae

Genus: Paracheirodon

Species: *innesi*

The neon tetra (*Paracheirodon innesi*) is a freshwater fish. The type species of its genus, it is native to blackwater or clearwater streams in southeastern Colombia, eastern Peru, and Western Brazil. The Neon Tetra was first imported from South America and was described by renowned ichthyologist G.S. Meyers in 1936, and named after Dr. William T Innes. *P. innesi*.

Biology

Its bright coloring is to make the fish visible in the dark blackwater streams, and is also the main reason for its popularity as a tropical fish. The neon tetra has a dark olive-green back over a silver-white abdomen. The fish is characterized by an iridescent blue-green horizontal stripe along each side of the fish from its nose to the base of the adipose fin, and an iridescent red stripe that begins at the middle of the body and extends posteriorly to the base of the caudal fin.

Size

It grows to approximately 3 cm (1.25 in) in overall length.

Sexual Dimorphism

The female having a slightly larger belly.

Neon tetras have a lifespan of about five years.

In the aquarium

Neon tetras are considered easy to keep in a community aquarium of at least 10 gallons, with a pH of 5.0–7.0 and KH of

1.0–2.0. They tend to be timid and, because of their small size, should not be kept with large or aggressive fish who may bully or simply eat them. Fish that mix well in an aquarium are other types of tetras, such as the rummy-nose tetra, cardinal tetra, and glowlight tetra, and other community fish that live well in an ideal Tetra water condition. Mid-level feeders, they are best kept in schools of five to eight or more, for the "shoaling" effect when they move around the tank. Their color and the iridescent stripe may become dim at night, and can be virtually invisible after a period of darkness. The color may also fade during a period of stress, such as human intervention into the tank. Neons are best kept in a densely planted tank with subdued light and an ideal temperature of 20–24°C (68–74°F).

Diet

Neon tetras are omnivores and will accept most flake foods, but should also have some small foods such as brine shrimp, daphnia, freeze-dried bloodworms, tubifex, and micro pellet food to supplement their diet.

Disease

Sporozoan Pleistophora hyphessobryconis causing disease Neon tetra disease.

Symptoms

Restlessness; fish begins to lose coloration; as cysts develop, the body may become lumpy; fish has difficulty swimming; in advanced cases the spine may become curved; secondary infections such as fin rot and bloating. To date, there is no known cure: the only 'treatment' is the immediate removal of diseased fish to preserve the remaining fish. The use of a diatom filter, which can reduce the number of free parasites in the water, may help.

Sexual Differences

The male is slender, and the blue line is straighter. The female is rounder, producing a bent blue line.

Reproduction

To breed Neon tetras, place a pair of the species in a breeding tank without any light, and gradually increase the lighting until spawning occurs. Other inducers include mosquito larvae and a hardness of less than 4 degrees. Some also recommend letting the level of nitrates rise, then do at least 50 per cent water change to simulate the fresh rain the tetras get in their natural habitat, the Amazon. It is recommended that everything you place in the aquarium be sterilized, as well as the aquarium top. Because the adults will often eat newly-hatched fry, it is best to remove them as soon as the eggs have been laid. The eggs are especially sensitive to light. Eggs will hatch within 30 hours of the laying. Fry can be fed rotifers, especially infusoria and egg yolk for 1 to 4 weeks, followed by nauplii of brine shrimp, shaved cattle liver, and formulated diets. Fry will achieve their adult coloration at approximately one month of age. Adults can spawn every two weeks.

5. Rosy Barb (*Puntius conchonius*)

Classification

Kingdom: Animalia

Phylum: Chordata

Class: Actinopterygii

Order: Cypriniformes

Family: Cyprinidae

Genus: Puntius

Species: *conchonius*

This fish was originally described by Francis Hamilton-Buchanan as *Cyprinus conchonius* in 1822 and is also referred to as *Puntius conchonius khagariansis* (Datta Munshi and Srivastava, 1988), *Systomus conchonius* or *Barbus conchonius*.

The rosy barb (*Puntius conchonius*) is a subtropical freshwater fish. It originates in Bengal, India and other parts of Southeast Asia.

Biology

This pinkish fish is one of the larger species of Barbs growing up to 6 inches (14 cm) in length. Their color becomes bolder during their mating periods. The male has a brighter pinkish color and the female is slightly plumper.

Habitat

In the wild their omnivorous diet consists of worms, insects, crustaceans, and plant matter. They have a lifespan of up to 5 years. Rosy Barbs natively live in lakes and fast flowing water in a subtropical climate. Their natural habitat has a pH of 6 to 8, a water hardness of 5-19 dGH, and a temperature range of 64–72 °F (18–22 °C).

Importance to Humans

The fish has commercial importance in the aquarium trade and is one of several species of barbs used to create hybrid versions of "tiger barbs".

Breeding

When the female is ready to spawn, she will appear swollen with eggs. The males will circle and chase the females, repeatedly nudging her head and belly area. Spawning usually occurs in the early morning, and lasts several hours resulting in several hundred eggs. Eggs are usually deposited in groups of plants, and the pair will attempt to eat any that they are able to locate. The young hatch in 24 to 36 hours, depending on water temperature. A day later, the young fish will hang on the plants, and/or the sides of the tank if the breeding takes place in an aquarium. In about six days the young are free-swimming and will seek out food. In captivity, they can be fed newly hatched brine shrimp.

Live Bearing Fishes

These fishes give birth to fully developed young fishes. Internal fertilization. Most of them are prolific breeders. They are cannibalistic, and they even consume their young ones. The most common types of livebearers are cultured in numerous color and finnage varieties. Commonly used species are as follows.

6. Guppy (*Poecilia reticultata*)

Classification

Kingdom: Animalia

Phylum: Chordata

Class: Actinopterygii

Order: Cyprinodontiformes

Family: Poeciliidae

Genus: Poecilia

Species: *reticulata*

Robert John Lechmere Guppy (1836-1916) discovered this tiny fish in Trinidad in 1866, although the fish was known to German aquarists prior to that time. With its original home in Central America, this most popular of all tropical aquarium fishes is very hardy and active. It is better to keep them in a large aquarium and in large numbers.

Biology

The guppy (*Poecilia reticulata*), also commonly known as guppie is one of the most popular freshwater aquarium fish species. The color of both male and female fish differs greatly. The tail pattern of the male fish also varies. Guppy is a prolific breeder and breeding occurs at interval of 4 to 6 weeks. It is a prolific mosquito larvae feeder and is used in the biological control of mosquitoes. Adult size 2.5 inches (6cm), lifespan 4 years, minimum tank size 5 gallon.

Diet

Omnivore, eats a variety of foods.

Water Quality

pH: 7.0-8.5, hardness:10-30dGH, temperature: 64-82 F (18-28ºC).

The female guppy is most often drab brown to grey in color. The somewhat smaller male has a naturally colorful caudal fin (tailfin).

Reproduction

The gestation period of a guppy is 22-26 days. When the female guppy becomes fertilized, a dark area near the anus, known as the gravid spot, will enlarge and darken. Guppies prefer water temperatures of about 22ºC (72ºF) for reproduction. After giving birth, the female is ready for conception within a few hours. If a male does not mate with the female soon after birth, the female can use stored sperm from the previous mating.

Adults will eat their young. Specially designed livebearer birthing tanks, which can be suspended inside the aquarium, are available from aquatic retailers. These serve the dual purpose of shielding the pregnant female from further attention from the males, and of providing a separate area for the newborn young as protection from being eaten by their mother. Another option for aquarists is to simply provide lots of plants for the young to hide in. Immediately after giving birth, female guppies experience a rush of hormones which serve to decrease their appetites. This can last up to 12 hours, which provides the fry ample time to hide provided that no other adult fishes are around. Guppies do not survive well in a closed environment.

7. Sailfin Molly (*Poecilia latipinna*)

Classification

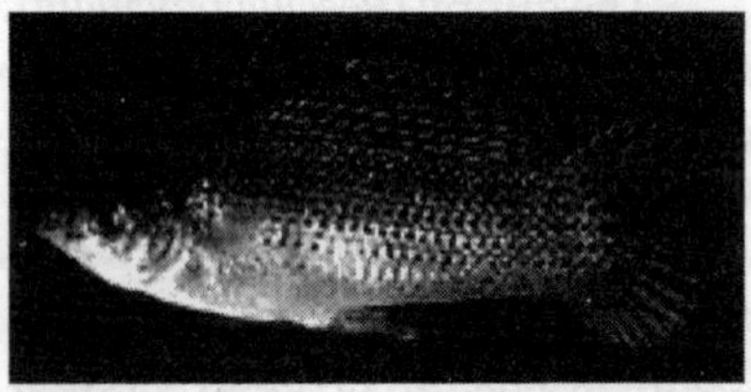

Kingdom: Animalia

Phylum: Chordata

Class: Actinopterygii

Order: Cyprinodontiformes

Family: Poeciliidae

Genus: Poecilia

Species: *latipinna*

The sailfin molly was originally described in 1821 by the naturalist Charles Alexander Lesueur. Lesueur based his description upon specimens from freshwater ponds in the vicinity of New Orleans, Louisiana.

Biology

Habitat

Sailfin mollies are most commonly observed in the shallow surface waters along the edges of marshes, lowland streams, ponds, sailfin mollies are able to survive oxygen depleted habitats.

Distinctive Features

The body of the sailfin molly is essentially oblong. The head is small and dorsally flattened, with a small, upturned mouth. The caudal peduncle is broad and the caudal fin is large, rounded, and sometimes tipped with black. The pelvic fins originate at a point anterior to the dorsal fin. In mature males the dorsal fin is greatly enlarged and colorful (it is this feature that gives the species its common name) and the caudal fin is similarly colorful, these conspicuous secondary sexual features play a role in female mate choices. Females tend to be larger and more plainly colored, a difference characteristic to the Poeciliidae. The body is generally light grey, although breeding males may be greenish-blue. Several rows of spots occur along the sides, back, and dorsal fin. Often these spots blend together forming stripes.

Size, Age, and Growth

The natural lifespan of sailfin mollies, males, which may live less than one year after achieving sexual maturity. At one year of age males typically range in size from 15-51 mm SL while mature females are likely to be approximately 19-53 mm The maximum recorded size for this species is 150 mm.

Food Habits

Sailfin mollies feed primarily upon algae and other plant materials, although they will consume a number of aquatic invertebrates including the larvae of mosquitoes.

Reproduction

Fertilization is internal and is accomplished by means of highly modified fin elements within the anal fin of males that form a structure known as the gonopodium. Sailfin mollies produce broods of 10-140 live young, depending upon maturity and size, and females may store sperm long after the demise of their relatively short-lived mates. The gestation period for this species is approximately 3-4 weeks, depending upon temperature, and a single female may give birth on multiple occasions throughout the year.

Predators

Sailfin mollies are members of the lower end of the food chain. As such they are prey for various animals including aquatic insects, other fishes, reptiles and amphibians, birds and mammals.

Parasites

The trematode, Saccocoelioides sogandaresi is a known parasite of the sailfin molly.

Importance to Humans

Artificially selected varieties are produced and sold in pet shops. Sailfin mollies may help to control mosquito populations by feeding on the larvae and pupae of these pests.

8. Green Sword Tail (*Xiphophorus hellerii*)

Classification

Kingdom: Animalia

Phylum: Chordata

Class: Actinopterygii

Order: Cyprinodontiformes

Family: Poeciliidae

Genus: Xiphophorus

Species: *hellerii*

Origin: Central America. Heckel, 1848 discovered this fish.

Biology

It is also called red swordtail. It is found in different colors like green, red and white. Some common varieties include wag tail, tuxedo and simpson.

Habitat

The green swordtail (*Xiphophorus hellerii*) is a species of freshwater. The green swordtail prefers swift-flowing, heavily-vegetated rivers and streams, but is also found in warm springs and canals.

Diet

Omnivorous, its diet includes both plants and small crustaceans, insects, and annelid worms.

Size

Male grows to a maximum overall length of 14cm (5.5in) and the female to 16 cm (6.3 in).

A live-bearer. The male fish has a large *sword* like structure which is the extension of the fin rays of the lower edge of the caudal fin. It uses its sword as a lever for jumping and as a rudder for steering. The name "swordtail" derives form the elongated lower lobe of the male's caudal fin (tailfin).

Sexual Dimorphism

Female being larger than the male but lacking the "sword". The male's "sword" is yellow, edged in black below.

Captive breeding. Female gives birth to as high as 200 young ones per litter. Sex reversal from female to male is common in this fish has a tendency of attacking smaller fishes.

It has caused ecological damage because of its ability to rapidly reproduce in high numbers.

9. *Pterophyllum scalare*

Classification

Kingdom: Animalia

Phylum: Chordata

Class: Actinopterygii

Order: Perciformes

Family: Cichlidae

Genus: Pterophyllum

Species: *scalare*

Biology

Pterophyllum is a small genus of freshwater fishes known to most aquarists as angelfish. All *Pterophyllum* species originate from the Amazon River basin in tropical South America. All *Pterophyllum* species form monogamous pairs. Eggs are generally laid on a flattened leaf or submerged log. The genus Pterophyllum discovered by Heckel, 1840.

Pterophyllum scalare this species discovered by, Lichtenstein, 1823. The best known species of angelfish is *Pterophyllum scalare*. Its natural color is silvery with three brownish vertical stripes. It is very peaceful (although sometimes can be aggressive to smaller fish, especially of its own breed) and is a popular aquarium fish. *P. scalare*, like all angelfish, comes from slow moving sections of rivers in the Amazon basin. Angelfish also do not inhabit dense vegetation except when very young, when they school together and hide in vegetation for protection. Adults are much more frequently found amongst sunken driftwood where very few plants grow. They spawn on broad-leaved sword plants in the wild, and prefer broad-leaved plants to spawn in an aquarium, if available.

Size

Maximum size is around 12–15 cm (up to 6 inches) length, up to 20 cm (8 in.) height.

Water

Prefer water with a 6.0–8.0 pH,with 6.5-7.4, water hardness of 5.0–13.0 dGH, and a temperature range of 24–30ºC (75–86ºF).

Average lifespan in an aquarium is 10 years. Freshwater angelfish are piscivores, meaning that they eat other fish. However, in an aquarium, their predilection for laziness sets in and they rapidly show preference for slow moving processed food rather than having to run down larger, harder to chew live fish. They are safe to keep with other peaceful fish that are not too small (*i.e.* those that could fit in its mouth). For example, an angel that has never seen a neon tetra might eat a small one just out of instinct if it can fit in its mouth.

10. Southern Platyfish (*Xiphophorus maculatus*)

Classification

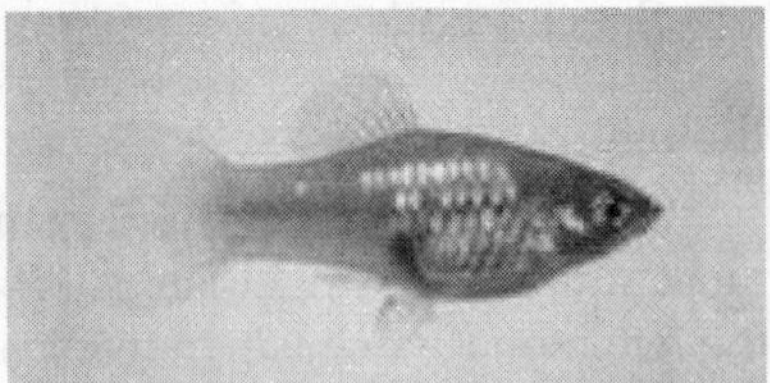

Kingdom: Animalia

Phylum: Chordata

Class: Actinopterygii

Order: Cyprinodontiformes

Family: Poeciliidae

Genus: Xiphophorus

Species: *maculatus*

The southern platyfish is commonly known simply as the platy (pl. platys or platies), from the fish's original generic name, *Platypoecilus*. Southern platyfish (*Xiphophorus maculatus*) is discovered by Günther, 1866. The southern platyfish (*Xiphophorus maculatus*) is a species of freshwater fish. It is native to an area of North and Central America stretching from Veracruz, Mexico, to northern Belize.

Biology

Habitat

X. maculatus prefers slow-moving waters of canals, ditches, and warm springs.

The southern platyfish grows to a maximum overall length of 6.0 cm (2.4 in). The life span 3 years.

Diet

Omnivorous, its diet includes both plants and small crustaceans, insects, and annelid worms.

Water Quality

pH 7.0–8.0, hardness of 9.0–19.0 dGH, and temperature range of 18–25 °C (64–77 °F). Minimum tank size 20 gallons.

Sexual Dimorphism

Male's modified anal fin (gonopodium) and caudal fin being more pointed, females grow larger and have a more rounded body.

Breeding

Platies are livebearers. Floating plants will help with the survival of the baby Platies, as the parents may turn cannibalistic. Breeding comes naturally to livebearers, so virtually no effort is needed to breed them. The babies should be removed as soon as they are free swimming to a separate tank to ensure their survival. In captivity, they reach maturity in 3-4 months and breed readily, the females giving birth to about 40–50 young a time.

Aquarium Plants

To provide some hardy aquatic plants in the aquarium, as they provide oxygen to the water, and shelter to the fish. Aquatic plants may serve as food, shade, or protection for fish, or they may support algae or small animals which are directly or indirectly food for game fish, or they may form habitats for the deposition of eggs.

The plants are usually arranged in groups, each group consisting of the same type. Each plant has to be fixed firmly deep in to the bottom layer of the sand, which can be done with the help of a "planting stick" or with fingers. After planting, the aquarium should be left undisturbed for about a week, before any fish is introduced into it. The following plants are suitable and commonly used in the aquarium:

1. *Vallisneria sprialis* (Eel-grass or Tape Grass)

Classification

Division: Spermatophyta

Sub-division: Angiospermae

Class: Monocotyledons

Series: Microspermae

Family: Hydrocharitaceae

Genus: Vallisneria

Species: *spiralis* L.

Morphology

Eelweed is a submerged grass-like perennial plant with ribbon-like leaves, 1 to 2 m long resembles a grass in general appearance. It grows well in quite freshwater, mostly in lakes or slow-moving streams, not more than 3 m deep.

Leaves are attached in clusters to a basal node of the rhizomes, up to 3 m long; and upper part of the leaf floating on the water surface. Leaves have netted veins; roots fibrous.

Stems buried in the mud, sending up tufts of ribbon like leaves in the water. Dioecious flowers are borne at the tip of a long peduncle with coils and retracts the fruit below the water after fertilization. Fertilization mechanisms is peculiar male flowers become detached

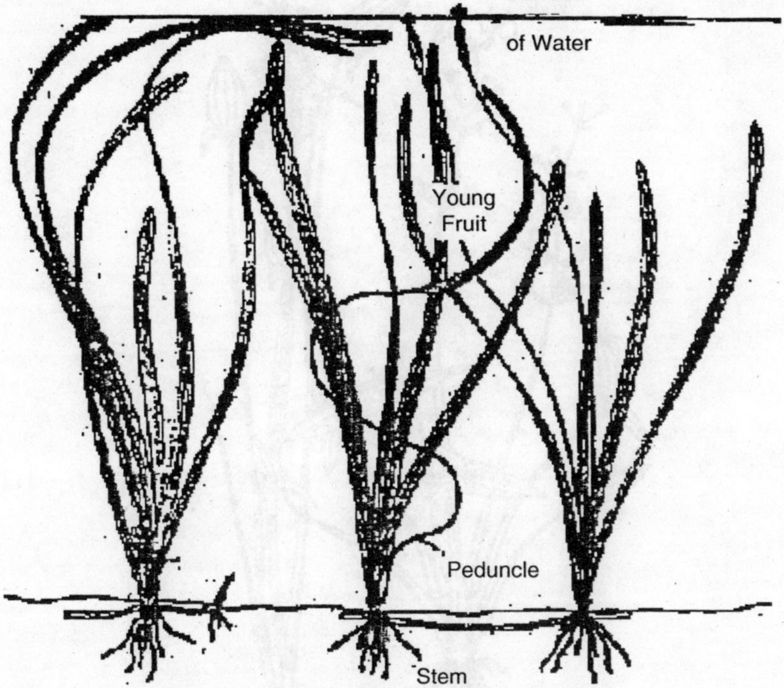

Vallisneria sprialis

from the plant and float to water surface and then fertilization takes place. Soon after the female flower goes down below the water surface and develop into berrylike fruits containing numerous seeds.

2. *Sagittaria guayanensis* (Arrowhead)

Classification

Division: Spermatophyta

Sub-division: Angiospermae

Class: Monocotyledons

Series: Apocarpae

Family: Alismaceae

Genus: Sagittaria

Species: *guayanensis* H. B. K.

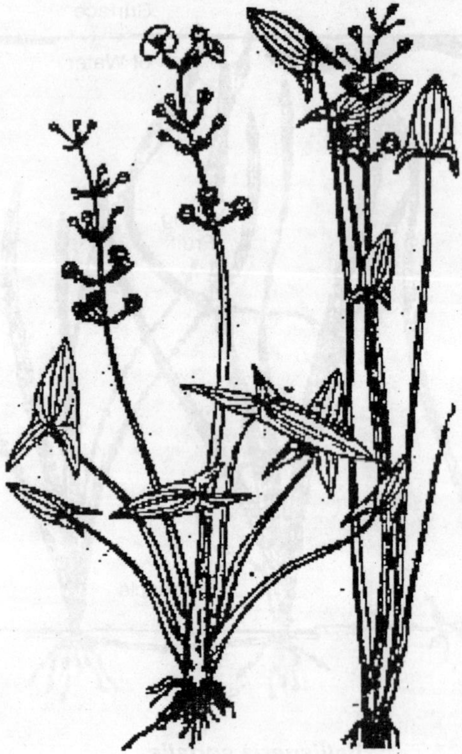

Sagittaria guayanensis

Morphology

Stems have milky juice; leaves erect, lance-shaped, 5-10 cm wide, natural look like a arrow head. Leaves grow from the rhizomes. White butter cup like flowers arise from the peduncle. Stamens and carpels in different flowers, the staminate flowers above. Leaves of 4 types, the arrow shaped and elliptical commonly immersed, and the tounge-like and ribbon-like submersed; two or more types may be found on the same plant. Bracts at the base of each flowers separate. Anthers about twice as long as wide. Tip of the leaf sharply angled Leaves 2 mm wide.

3. *Cabomba caroliniana* (Fanwort)

Morphology

C. caroliniana Gray. Stem slender, with a gelatinous slime; leaves opposite or whorled, the thread-like division branching in fan-like fashion, superficially resembling those of Coontail; flowers about 1cm. long.

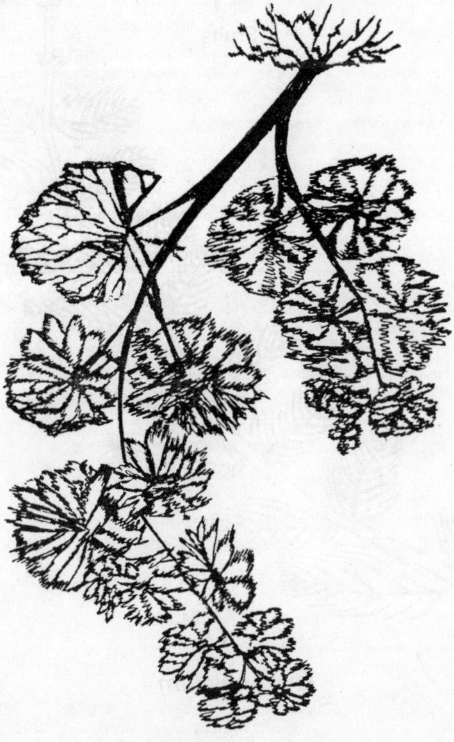

Cabomba caroliniana

4. *Myriophyllum* (Water Milfoil)

Morphology

There are species *M. exalbescens*, *M. heterophyllum*, *M. humile* and *M. scabratum*.

Calyx lobes 4; petals 4 or none; stamens 4 or 8; fruits of 4 joined nut-like bodies. With a few exceptions, the species look much alike, and some are nearly impossible to distinguish without flowers or

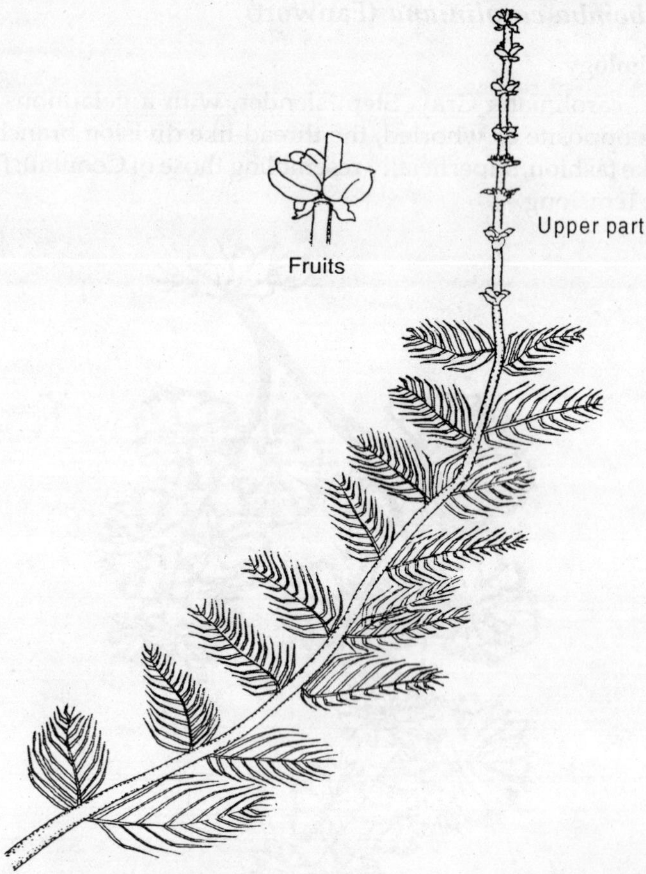

Myriophyllum

fruits. Leaves cut into thread like or narrow ribbon-like. Leaves small, roundish. Leaves 5-12 mm. long. Flowers in spikes borne above the water.

Floral bracts mostly scattered.

5. *Ceratophyllum demersum* (Coontail)

Classification

Division: Spermatophyta

Sub-division: Angiospermae

Class: Dicotyledons

Series: Ordines

Family: Ceratophyllaceae

Genus: Ceratophyllum

Species: *demersum* L.

Morphology

C. demersum L. Plants entirely submersed, without roots, appearing olive-green when seen through the water; leaves in whorls,

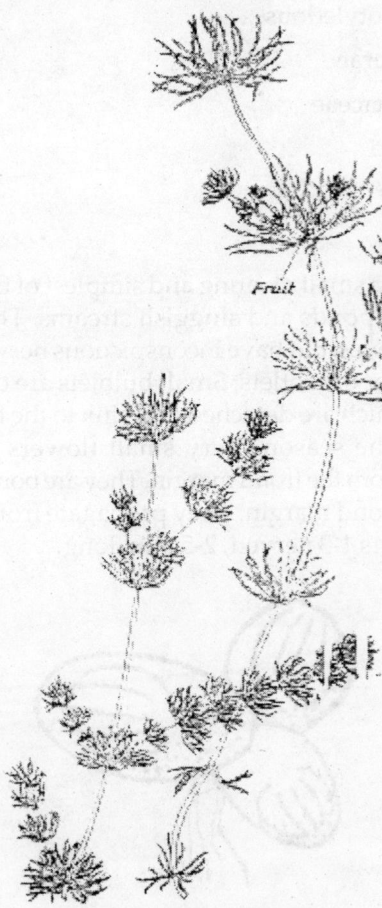

Ceratophyllum demersum

stiff, repeatedly forked, the thread-like divisions with teeth along one side; leaves more crowded toward the tip, giving the "coontail" appearance; fruits nut-like, borne without stalks in the leaf axils. Widely distributed, usually in hard water. Very variable in length and degree of crowding of leaves.

6. *Lemna minor* (Duck weed)

Classification

Division: Spermatophyta

Sub-division: Angiospermae

Class: Monocotyledons

Series: Nuliflorae

Family: Lamnaceae

Genus: Lemna

Species: *minor*

Morphology

Duckweed is a small, floating and simplest of flowering plants, found in stagnant ponds and sluggish streams. They are stemless and lack of true leaves; but have inconspicuous nerves, thick rounds fronds and number of rootlets. Small bulblets are developed at the edge of the leaf which are detached and sink to the bottom and then germinate after the season. Very small flowers are borne on a peduncle arising from the frond margin. They are borne on a peduncle arising from the frond margin. They propagate from the base of the parent frond. Plants 1-3 nerved, 2-5 mm. long.

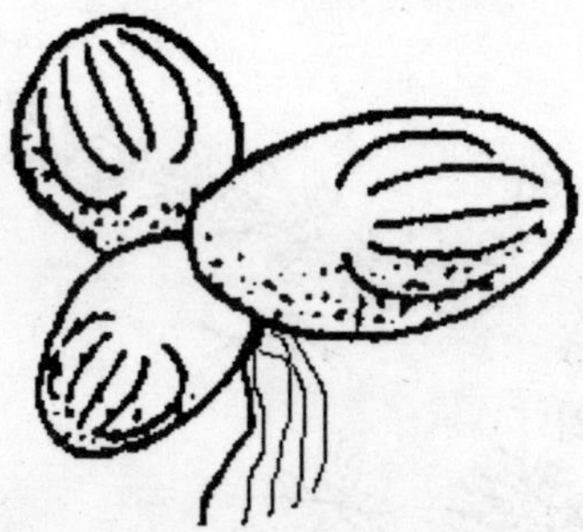

Lemna minor

7. *Hydrilla verticillata*

Classification

Division: Spermatophyta

Sub-division: Angiospermae

Class: Monocotyledons

Series: Microspermaceae

Family: Hydrocharitaceae

Genus: Hydrilla

Species: *verticillata*

Morphology

Hydrilla is a submersed aquatic weed rooted to the bottom forming large masses, leafy, dioecious. Since its light requirement is extremely low, it grows in deep water up to 10 to 12 m. it is found in drainage channels, canals, ponds and lakes. Stems are long-branching which may break and form floating mats and grow very fast. The lower leaves are opposite and small; while the upper ones are in whorls and longer. Flowers arise from the spathe singly near the water surface on the growing tip; ovary extending beyond the spathe into a beak. Reproduction is largely vegetative. It is most

Hydrilla verticillata

often used as egg collector of freshwater fish carps; which helps to collect eggs easily along with hydrilla and transfer to hatchery.

8. *Ludwigia* (False Loosestrife)

Classification

Division: Spermatophyta

Sub-division: Angiospermae

Class: Dicotyledons

Series: Calyciflorae

Family: Onagraceae

Genus: *Ludwigia* L.

Morphology

Leaves opposite, alternate, without petioles, narrowed to a petiole. Flowers not stalked; petals small or absent. Plants nearly

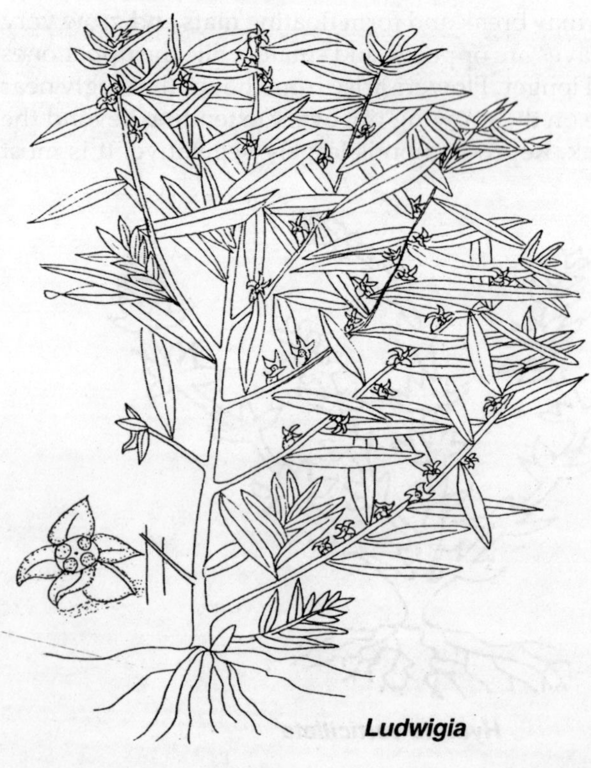

Ludwigia

prostrate on the mud or in shallow water. Flowers 12-13 mm. broad, 5-12 mm. long. Fruit several times as long as thick. Fruit nearly globular. Fruit tapered from apex to base.

Chapter 25

Design, Construction and Materials Used in Home and Public Aquaria

Construction of an Aquarium

The construction of an aquarium is a skillful job and depending on the requirement of hobbyist, it can be prepared/constructed. Size also play important role in deciding the thickness of glass used for construction and also the angle size. In recent years, all-glass aquaria have become more popular these are prepared by using only glass but for rough handling and considering durability, aquaria with frame are good.

For constructing the aquaria many things are needed but some essential requirements are as follows.

1. Aluminium angles of different sizes
2. Hack-saw blade
3. Ribbits/screw
4. Glass with variable thickness (5 mm, 8 mm, 10 mm, 12 mm and 20 mm)
5. Diamond glass cutter (hirkani)
6. Bowl
7. Stove
8. Clamp
9. Driller

10. File
11. Hammer
12. Anvil or metallic cube
13. Butamin solution (tar) silicon sealant tube
14. Gun etc.

Setting of an Aquarium

Various sizes of aquarium is available in market. An aquarium can be of a small size to occupy the limited space available in the room, or of a fairly large size. However, it is easy to look after and maintain a small aquarium, which may be 45 x 30 x 20 cm or 60 x 30 x 38 cm, and can be purchased from the market or prepared at small cost.

The aquarium is made of iron or aluminum frame with glass plates of suitable thickness depending up on its size. The glass plates are fixed to the frame by means of aquarium cement or a mixture of putty and red lead. The bottom plate can be 5–6 mm thick made of wired glass, while the side panels are made of transparent thick glass. The metallic frame can be painted green or red and the aquarium is allowed to dry for a few days and tested for being leak proof.

The aquarium is placed on a table near a window so that it receives sufficient light. If this is not possible, the aquarium is provided with a cover having an electric bulb or fluorescent tube. A bulb of 60 watt is sufficient to provide light for a medium sized aquarium, but it should be placed at a distance from the water surface to avoid may heating up of the water. An aquarium can also be made to rest on a thick wooden support firmly fixed to the wall. Generally, provision of artificial light for illumination of the tank for a short period is necessary.

The first step for setting the aquarium is to cover the bottom with a layer of clean sand (about 2.5 cm thick). A mixture of coarse sand and leaf mould is then spred over the bottom to form 2.5 cm thick layer. This along with the fish excreta forms good manure for the plants. The mixture is covered with 1.2 cm thick layer of washed sand. Thick "morang" can also be used in place of sand, but it must be washed thoroughly before being placed in the aquarium. Small

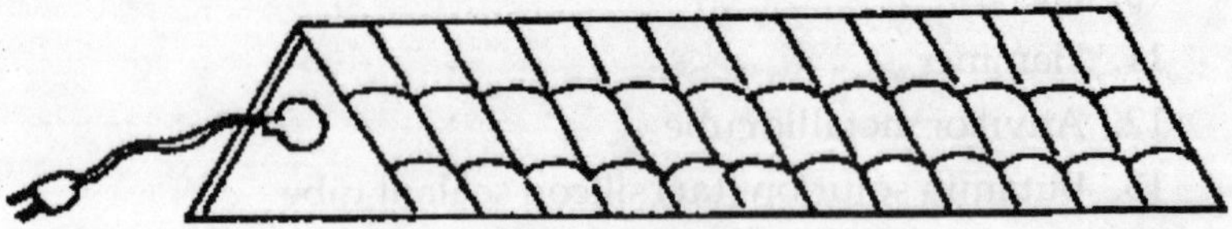

Figure 25.1: Aquarium Hood

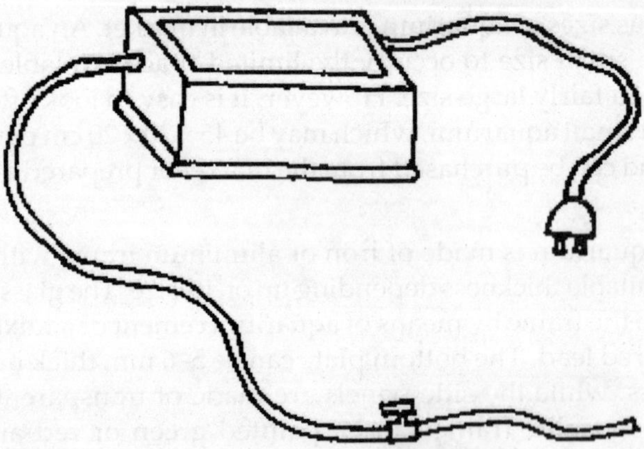

Figure 25.2: Aerator

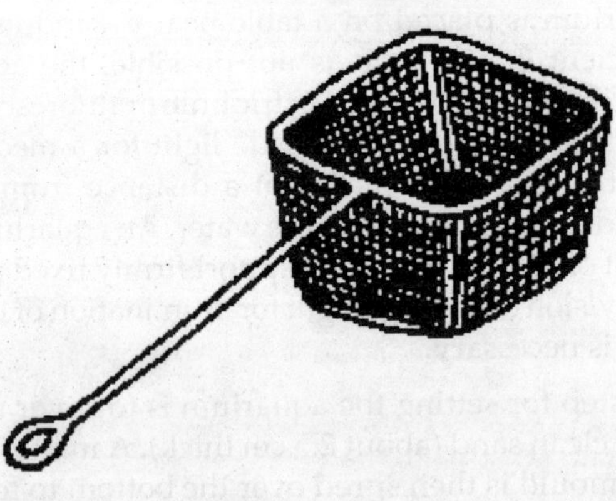

Figure 25.3: Hand Net

rocks, colored stones, shell, corals, or any other thing which looks attractive can be placed on the top of the sand.

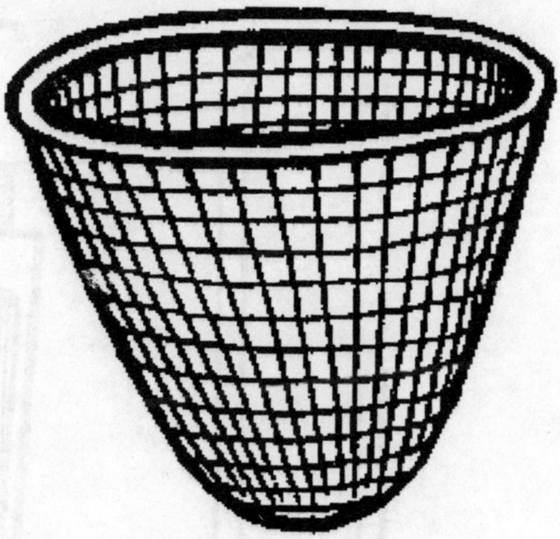

Figure 25.4: Worm Feeder

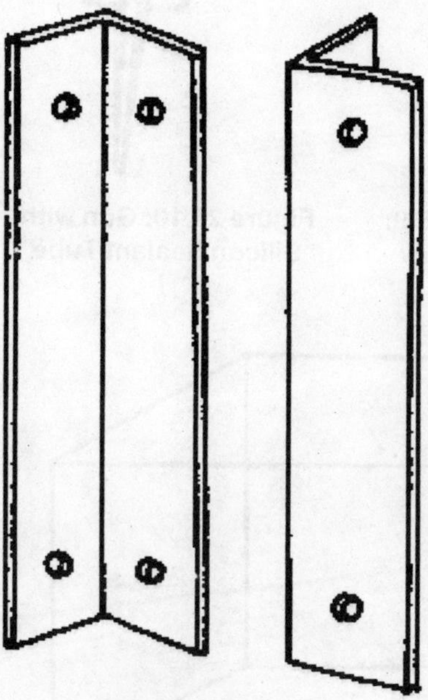

Figure 25.5: Aluminium Angles

Figure 25.6: Glass Cutter

Figure 25.7: Driller

Figure 25.8: File

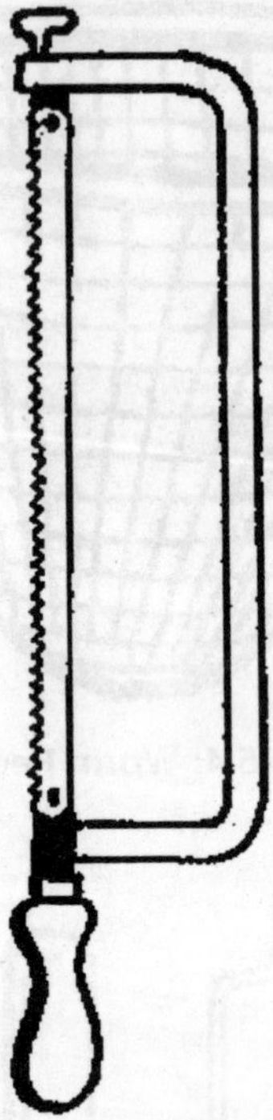

Figure 25.9: Hacksaw

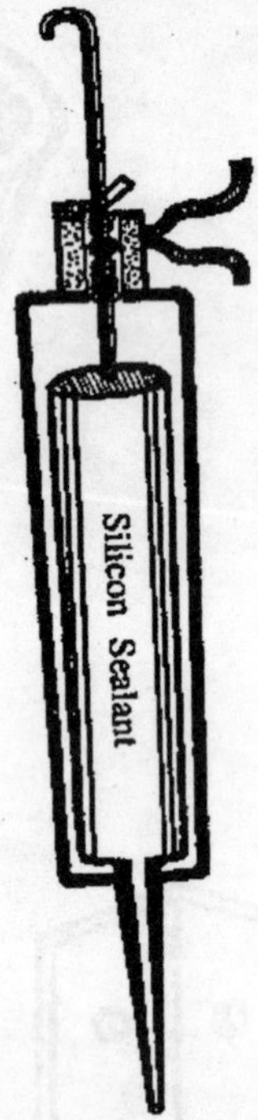

Figure 25.10: Gun with Silicon Sealant Tube

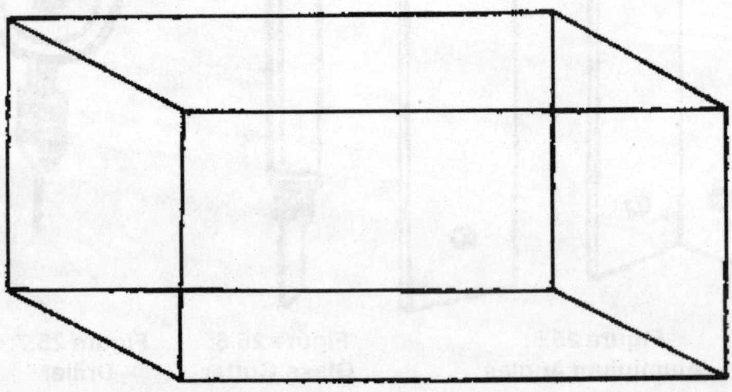

Figure 25.11: All Glass Aquarium

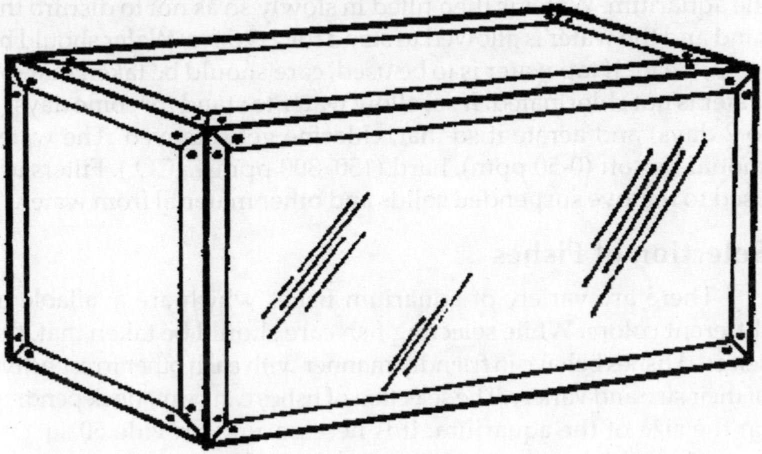

Figure 25.12: Framed Aquarium

Figure 25.13: Wellset Aquarium

Sand, gravels to create / stimulate the natural environment, it also gives support for fixing aquatic plants. The faecal matter will not be directly visible and get settled into cervices and the beauty of aquarium will not get affected. Sand or gravels before using it should be thoroughly washed with water to remove dust and dirt. A background can be provided for the aquarium by suitable painting of the back panel, or by placing a soft board painted in colors, behind the aquarium. Water is then filled in slowly so as not to disturb the sand and the water is allowed to stand for 2-3 days. Water should be fresh, clean. If tap water is to be used, care should be taken that the water is not chlorinated. If so, allow water to stand for some days (1 to 2 days) and aerate it so that, chlorine get removed. The water should be soft (0-50 ppm), hard (150–300 ppm $CaCO_3$). Filters are used to remove suspended solids and other material from water.

Selection of Fishes

There are variety of aquarium fishes which are available in different colors. While selecting fish care should be taken that, the selected fishes behave in friendly manner with each other irrespective of their size and variety. The selection of fishes can be made depending on the size of the aquarium. It is necessary to provide 50 sq. cm. water for a fish having size 2.5 cm. long. In a 2′ x 1′ x 1′ aquarium we can keep 36 fishes of size 2.5 cm long. Select fishes of different varieties for better look and liveliness of an aquarium. Select healthy fishes.

Selection of Plants

The aquarium plants enhance the beauty of aquarium, also the utilize CO_2 and release O_2. They maintain the equilibrium of the aquarium. They also provide place to hide for small fishes and breeding beds for some fishes. Depend upon the size of aquarium plants selected.

Food for Aquarium Fishes

Fishes have to be fed regularly with suitable food. Dry fish food can be obtained from the market but over feeding is to be avoided as it causes sickness and death. Fish should not be fed on the same diet for a long period and food should be changed from time to time. Hence fish should be given dry food and live food alternately. Water fleas and tubifex can be obtained from a lake or pond by means of a small net, and constitute good live food for the fish. Minced meat can be given occasionally. Leafy green vegetables, lightly boiled and minced, can also be given for a change. Eggs and larvae of mosquito, shrimp, earthworm and crustacean larvae collected from the pond can also be given as live food.

For live food use worm feeder basket. Basket float on water, fish collect the larvae from small holes of basket. Due to that no wastage of food unused food remains in the basket.

Maintenance

An aquarium, if prepared carefully, does not need any special care, but cannot be left unattended and some precaution are necessary. Large fluctuation of the water temperature should be checked and a thermometer can be adjusted for keeping a watch. The preferred temperature is 22-25°C.

Aerator

An electrically operated aerator is a useful accessory for maintenance of the aquarium. It is used to generate a continuous stream of air bubbles using a "diffuser". Thus, water circulates from bottom to the surface and is oxygenated from time to time. These aerators drive out carbon dioxide from water.

Heater

A thermostatically controlled water heater can be fitted in the aquarium, which automatically maintains water temperature within the desired range during winter season. Excess of food and waste matter must be removed by means of a siphon. A long glass tube can be used to siphon out waste and undesirable matter from time to time.

Removal of Algae

A green or brown film of algae appears on the inside of the glass panels, and must be removed from the front and side glasses. Green algae on the back panel and stones can be left as such as it serves to oxygenate water. To remove the algae, an ordinary razor blade can be fixed at the end of a wooden rod to scrap and clean the glass.

When water level in the aquarium falls due to evaporation, freshwater is added to make up the loss. The aquarium is emptied ½ or 2/3rd once in 2-3 months, the glass panels are cleaned, and freshwater is added to fill it up to the required level. Filters are used in an aquarium to clear its water.

If the aquarium does not get sunlight, the artificial light from the bulb must be given for 6-8 hours every day.

Hand Nets

It is a rectangular in shape made up from small meshed mosquito cloth. Aquarium fish should be handled with the help of a hand net.

Any diseased fish should be immediately removed from the aquarium. With these precautions the aquarium can be maintained for a fairly long time. The aquarium can be disinfected by soaking for 24 hours in every strong potassium permanganate, or by adding 'Dettol' until the water turns milky. In both the instances the plants and gravel should be removed and treated separately. The alternative is to remove all fish from the aquarium for about four weeks, in which time the disease or parasite should have been eliminated.

Chapter 26
Common Diseases of Aquarium Fishes and their Diagnosis and Treatment

A large number of diseases occur in all aquatic animals including fish, due to nutritional deficiency or unhygienic condition of water or attack of parasites. All species fish become infested with pathogenic bacteria, protozoans, fungi, worms, helminthes and crustaceans causing sickness. Non-communicable diseases are caused by nutritional, environmental and genetic factors, besides neoplastic diseases such as, tumors and cancers. To maintain a healthy aquarium these diseases need to be prevented, identified and controlled. Some of the measures which are taken to control the diseases are disinfection with dilute potassium permanganate solution, maintenance of optimum levels of temperature, light, pH and oxygen. Moreover, disinfection of equipments used in an aquarium is necessary.

Of prime importance is the strict observation of quarantine and the disinfection of all new stock–both fish and plant.

Diagnosis

By carefully checking the symptoms of the affiliated fish against the following list, the aquarist will be able to diagnose the most probable cause of the malady. By referring to the specified complaint in the following pages, the fish keeper will find both the cause of the problem, together with the treatment. It should hardly need saying that, not only must action be taken to cure the complaint, but steps must also be taken to remedy the original cause of the problem. The list is:

1) White cotton woolly tufts on the body or fins-*fungus*. 2) White film on the eye-*Cataract*. 3) Fish knocks against firm object and has a flat transparent green creature, about the size of a ladybird, attached to it-*Argulus*. 4) Fish rubs and scratches itself, fins twitch, and the fish makes sudden will dashes, breathing rate accelerated-*Flukes*. 5) Gill inflamed and swollen, skin dull with milky-white patches on the body, loss of appetite-*Branchitis*. 6) Scales stand out, bloated body-*Dropsy*. 7) White pin-head spots on body and fins-*Ichthyophthiriasis*. 8) Thread like worm attached to fish-*Lernaea*. 9) Difficulty in swimming. Fish cannot rise from bottom. Floats like a cork at water surface. Turns on back. Floats head down- *Disarrangement of air-bladder*. 10) Air bubbles in faces, the fish is sluggish with swollen belly-*Indigestion*. 11) Slow clumsy movements, fish remains stationary and waddles body from side to side- *Shimmies*. 12) Fish hollow-eyed with sunken chest, loss of appetite and sluggish-*Tuberculosis*. 13) Air-bubbles in fins. Air Embolism. 14) Blood-streaked veins in caudal and other fins. Later the fins split and rot- *Fin-rot*. 15) Eyes protrude on normal-eyed fish-*Exophthalmis*. 16) Dark grey patches on body which later turn black and peel leaving raw spots-*Melanosis*. 17) Small worm-like creatures attached to fish-*Leeches*. 18) Slimy, greyish-white mucous covering body. Fins folded, loss of appetite. Fish may lie on its side. Most obviously noticeable on dark colored fish-*Chilodonelliasis*. 19) Pale gills with signs of rotting. Breathing rate much increased. Loss of appetite-*Gill rot*. 20) Cotton wool tufts at mouth and on gills. Swollen lips. Loss of appetite. Sluggish–*Mouth fungus*. 21) Lies near surface without movement. Fish lies on its side. Gills inflamed-*Red Pest*. 22) Open sores on the body-*Ulcers*. 23) Lack of appetite and swollen belly-*Constipation*.

Diseases and their Treatment

Fungus (Saprolegnia)

This cotton wolly fungus attacks only fish which have been wounded, weakened by parasites, or kept in bad conditions. The fungus spores are present in most waters, but cannot infect healthy fish. The individual threads of the fungus grow into the skin like roots. Remedy the conditions which have caused the infection.

Treatment

If only small areas are affected by fungus, touch the infection with iodine solution. First iodine dilute by the addition of an equal amount of water, then apply. If large areas are covered treat with salt or Phenoxethol. Phenoxethol 10 per cent, 90 ml solution per gallon of water for treatment. Infected fish can remain in the bath until cured.

Cataract

There are two forms:

Cataracta Traumatica

Thus disease is characterized by a whitish film covering the lens of the eye. Generally only one eye is affected. Cause of disease not yet known.

Treatment

The iodine treatment (iodine dilute with equal volume of water), to which nine parts of glycerine has been added, will often bring about a cure.

Cataracta Parisitica

At first glance resembles the first form. Closer inspection through a magnifying glass will reveal the film to be composed of white spots. In reality the dots are little worms, which feed upon the lens and destroy it. Worm cataracts are due to the larvae of sucking worms. The adult parasite lives in water-birds, eggs are discharged with the droppings into the water. The eggs hatch in the water, the resulting miracidiae being taken into the body of a snail, where they change shape and become sporocysts. After about six weeks the larvae leave the liver of the snail and move into the water. Upon meeting a fish the larvae enter the body and migrate to the eye. If the larvae fail to penetrate the fish within two days they die.

Treatment

Treat the affected fish with Phenoxethol (10 per cent solution), also soak food in the solution prior to feeding. If the eye is too badly affected it may not be possible to cure it–destroy the fish.

Argulus (Fish louse)

One of the better known fish parasites. A round, flattish creature armed with two large suckers and eight legs. In addition it has a large hallow 'sting'. The 'sting' is pushed under the scale of the fish to feed upon the blood of its hosts. The louse can be introduced to the aquarium or pond with Daphnia, so care should therefore be taken when screening this food.

Treatment

Being a large creature, the louse can be removed with forceps or, alternatively, they can be rubbed off. Hold the fish in a wet net when carrying out these operations. Allow the tank to dry out for 7 days, to kill both eggs and adults. If preferred the potassium permanganate (1 per cent) 2ml per gallon of water for bath treatment.

Gill Fluke (Dactylogyrus)

Infects the gills of the fish and are egg-laying. Fish affected by these parasites have increased breathing frequency, the gills are stretched wide open and pale in color. The eggs are very resistant to chemical treatment, and for this reason it is necessary to make repeat treatments to kill the newly hatched flukes.

Treatment

Treatment may be given by first applying the 15 minute ammonia bath (10 per cent, 45 ml per gallon of water), bath not more than 20 minutes, afterwards placing the fish in a long-term bath of Methylene Blue (1 per cent, 4ml per gallon of water).

Skin Fluke (Gyrodactylus)

Closely related to the former fluke, it is live-bearing. Fish affected by this fluke will tend to lose some color, the fins drop and may become torn, the skin becomes very slimy and may show small blood spots. In strong light it may be possible to see these tiny parasites, especially if the fish has a dark color. The fluke is an unsegmented flat worm, with an elliptical shaped body that is flattened on the ventral side. At the front there are two conical projections, through which glands pass a thick, sticky liquid with which it clings to the fish. The other end carries a strong disc-shaped attachment of small hooks in the shape of a ring in the middle of which are two larger hooks. The parasite first grips the skin of the fish with the small hooks and then drives in the large ones.

Flukes can be introduced by anything that has been in fluke infested waters. Disinfection and quarantine techniques will do much to avoid the problem.

Treatment

Treatment of affected fish is as mentioned for the elimination of the Gill Fluke. 'Sterazin', a commercial preparation, is possibly the best of all treatments.

Bronchitis (Inflammation of the Gills)

This is caused by polluted water or over-crowding reducing the oxygen content of the water.

Treatment

Remedy the cause and give the fish treatment in methylene blue (1 per cent, 4ml per gallon of water).

Dropsy

Is an accumulation of fluid within the tissues of one or other of the intestinal organs. The belly of the fish swells so greatly that it appears likely to burst. However, the fish dies long before the pressure becomes great enough for this to happen. The belly is filled with an almost watery liquid. The disease can be infectious, and is caused by a bacterium which can always be found in water and milk. Investigations seem to indicate that the disease mainly attacks fish that are poorly fed and have lowered resistance.

Treatment

Treat the fish with chloromycetin (use 60 mg. to each gallon of water for 4 days, or 600 milligrams per gallon for 24 hours. Treatment should not be extended beyond the stated periods), keep the water cool and do not feed. Difficult to cure.

Lernea (Anchor Worm)

Only the female of this parasite attacks fish. She has a long body with an anchor-like appendage on the head. The creature is not, in fact, a worm but is related to the Cyclops. It grows to a length of around ¾ inch (19mm). By means of the anchor-like appendages, which are buried deep into the muscles of the fish, the creature become very firmly attached to her unwilling host. After shedding her fertilized eggs the female dies, leaving deep wounds in the fish which heal very slowly. The young which hatch from the eggs can re-infest the fish.

Treatment

To rid fish of this parasite it must be held in a wet net, and the parasite touched with a brush dipped in very strong potassium permanganate. This should kill the parasite so that it can be removed with forceps. Touch the wound with iodine solution (first iodine dilute by the addition of an equal amount of water, then apply) and cover with Friar's Balsam. The treated fish must be placed in clean water, while their infected quarters are kept free of fish for a period of about two months, during which time the pests, young and adult, should have died through lack of a host.

White Spot (Ichthyophthirius)

This parasite is a protozoan which penetrates the upper layer of the epidermis and causes small whitish blister to appear. The parasite feeds upon the red blood corpuscles of the fish. When mature, the parasites leaves its host and sinks to the bottom, where is encysts in a jelly-like covering. Rapid division takes place within the cyst to produce between 500 and 1000 young parasites. At a temperature of 68°F (20°C), the process takes about 18 hours. The rate at which the parasite infection increases can so weaken the fish that it will die unless treated. If the young parasite does not find a host within a few days it dies. White spot is introduced by infected fish and/or plants and material that has been in contact with infected fish.

Treatment

A strict routine of disinfection and quarantine will prevent the parasites infecting healthy stock. To be quite sure that newly acquired goldfish are free of the parasite, they should be kept in isolation for at least four weeks. Treat in a permanent bath of malachite green (prepare a stock solution by adding 1 gram to 500 cc of distilled water. The medicinal bath is made by mixing 9 cc of the stock solution into one gallon of water. Do not use in zinc or galvanized iron containers. Immerse the fish for up to 1 hour) or methylene blue (1 per cent, 4ml per gallon of water).

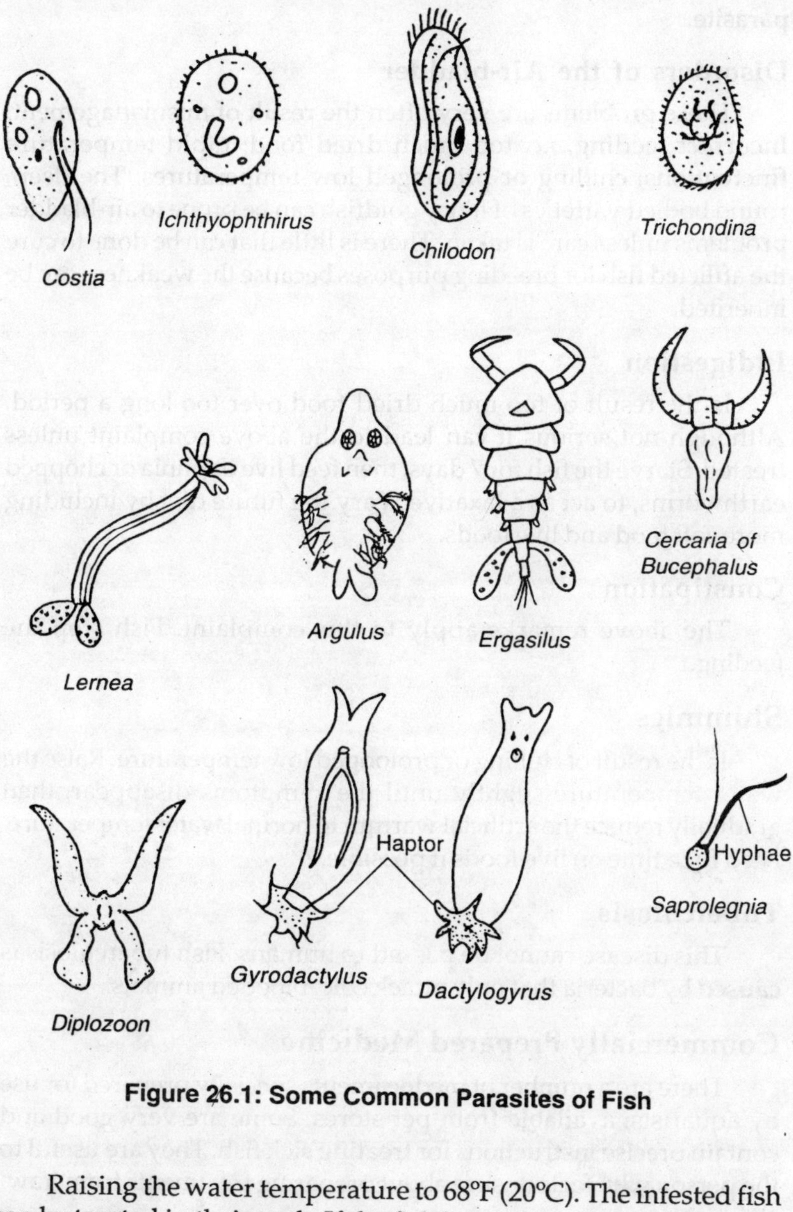

Figure 26.1: Some Common Parasites of Fish

Raising the water temperature to 68ºF (20ºC). The infested fish can be treated in their tank. If the fish lives in a pond, the parasites will die after two months if every fish is removed. There are some excellent commercial remedies for the eradication of this dangerous parasite.

Disorders of the Air-bladder

These problems are very often the result of mismanagement. Incorrect feeding, *i.e.* too much dried food; rapid temperature fluctuations; chilling or prolonged low temperatures. The deep, round bodied varieties of fancy goldfish can be prone to air-bladder problems unless care is taken. There is little that can be done to cure the afflicted fish for breeding purposes because the weakness can be inherited.

Indigestion

Is the result of too much dried food over too long a period. Although not serious it can lead to the above complaint unless treated. Starve the fish for 7 days, then feed live daphnia or chopped earthworms, to act as a laxative. Vary the future diet by including more soft food and live foods.

Constipation

The above remarks apply to this complaint. Fish stop the feeding.

Shimmies

Is the result of chilling or prolonged low temperature. Raise the water temperature slightly until the symptoms disappear, then gradually reduce the artificial warmth to normal water temperature. Feed for a time on live foods if possible.

Tuberculosis

This disease cannot be passed to humans. Fish tuberculosis is caused by bacteria that only attack cold-blooded animals.

Commercially Prepared Medicine

There are a number of medocaments specially prepared for use by aquarists, available from pet stores. Some are very good and contain precise instructions for treating sick fish. They are useful to the person who feels unsure about preparing treatments from 'raw' chemicals.

Bibliography

Adhikari, Subhendu & D K Chatterjee. Management of Tropical Freshwater Ponds. Daya Publishing House, New Delhi.

Agarwal, S C. History of Indian Fishery. Daya Publishing House, New Delhi.

Ahilan, B *et al.* Textbook of Aquariculture. Daya Publishing House, New Delhi.

Andrew, P. Economics of Brackishwater Shrimp Culture. Daya Publishing House, New Delhi.

Anmarie Barrie. Our First Aquarium.

Balachandran, K K. Postharvest Technology of Fish and Fish Products. Daya Publishing House, New Delhi.

Baluyut, Elvira A. Aquaculture Systems and Practices: A Selected Review/FAO. Daya Publishing House, New Delhi.

Belsare, D.K. Tropical Fish Farming. Environmental Publication, Karad.

Bensam, P. Development of Marine Fisheries Science in India. Daya Publishing House, New Delhi.

Biswas, K P. Corals of Tropical Oceans. Daya Publishing House, New Delhi.

Biswas, K P. Economics in Commercial Fisheries. Daya Publishing House, New Delhi.

Biswas, K P. Harvesting Aquatic Resources. Daya Publishing House, New Delhi.

Biswas, K P. Industrial Fisheries. Daya Publishing House, New Delhi.

Cadima, Emygdio L. Fish Stock Assessment Manual/FAO. Daya Publishing House, New Delhi.

Chakraborty, Chiranjib & A K Sadhu. Biology Hatchery and Culture Technology of Tiger Prawn and Giant Freshwater Prawn. Daya Publishing House, New Delhi.

Chakraborty, N.M. Biology Culture and Production of Indian Major Carps: A Review. Narendra Publishing House, New Delhi.

Chattopadhyay, G N. Chemical Analysis of Fish Pond Soil and Water. Daya Publishing House, New Delhi.

Chaudhuri, A B. Biodiversity of Mangroves. Daya Publishing House, New Delhi.

Chaudhuri, A B. Endangered Wetlands. Daya Publishing House, New Delhi.

Chaudhuri, Hiralal & A B Chaudhuri eds. Aquaculture Beyond 2000: New Horizons. Daya Publishing House, New Delhi.

Chaudhuri, Hiralal & A B Chaudhuri eds. Aquaculture Ichthyology and Limnology. Daya Publishing House, New Delhi.

Chauhan, R S. Gramin Kshetro mein Samanvit Battakh Sah Matsaya Palan. Daya Publishing House, New Delhi.

Chauhan, R S. Matsaya Paalan Digdarshika. Daya Publishing House, New Delhi.

Chonder, S.L. Induced Carp Breeding.

Coche, A G & J F Muir. Meetha Jal Machali Palan ke Liye Talab Nirman/FAO. Daya Publishing House, New Delhi.

Coche, A G & J F Muir. Pond Construction for Freshwater Fish Culture: Pond Farm Structures and Layouts/FAO. Daya Publishing House, New Delhi.

Cuvier, Baron. Handbook of Mollusca and Radiata. Daya Publishing House, New Delhi.

Dholakia, A D. Fisheries and Aquatic Resources of India. Daya Publishing House, New Delhi.

Dholakia, Anshuman D. Ornamental Fish Culture and Aquarium Management. Daya Publishing House, New Delhi.

FAO. Handbook on Small Scale Freshwater Fish Farming. Daya Publishing House, New Delhi.

FAO. Integrated Agriculture Aquaculture: A Primer. Daya Publishing House, New Delhi.

Fifth Indian Fisheries Forum, 2000. Central Institute of Fresh Water Aquaculture, Bhubaneswar, Orissa.

Gahlawat, S K *et al.* Manual of Experimental Ichthyology. Daya Publishing House, New Delhi.

Gautam, Ashutosh & N K Agarwal eds. Recent Researches in Aquatic Environment. Daya Publishing House, New Delhi.

Gautam, Ashutosh ed. Conservation and Management of Aquatic Resources. Daya Publishing House, New Delhi.

Giriappa, Somu ed. Role of Fisheries in Rural Development. Daya Publishing House, New Delhi.

Gonzalez, J F. Matsiyaki Udhyog ķe Apjal ka Upachaar / FAO. Daya Publishing House, New Delhi.

Gonzalez, J F. Wastewater Treatment in the Fishery Industry / FAO. Daya Publishing House, New Delhi.

Gupta, R K *et al.* Fundamentals of Ichthyology. Daya Publishing House, New Delhi.

Hameed, M Shahul & M R Boopendranath. Modern Fishing Gear Technology. Daya Publishing House, New Delhi.

Hiware, C.J. and S.R. Sonwane. Hand Book of Fish Aquarium.

Hosetti, B B & Arvind Kumar. A Textbook of Applied Aquatic Biology. Daya Publishing House, New Delhi.

Humphrey, John et al. Aquatic Animal Quarantine and Health Certification in Asia / FAO. Daya Publishing House, New Delhi.

Insull, David & Colin E Nash. Aquaculture Project Formulation / FAO. Daya Publishing House, New Delhi.

Jhingran, V.G. Fish and Fisheries of India. Hindustan Publishing Corporation, New Delhi.

Jyoti, M K & Arti Sharma. Fishes: Aid to Collection Preservation and Identification. Daya Publishing House, New Delhi.

Kar, Devashish. Fundamentals of Limnology and Aquaculture Biotechnology. Daya Publishing House, New Delhi.

Khanna, D R & S Singh. Pond Fish Ecology and Economics. Daya Publishing House, New Delhi.

Khanna, D R *et al* eds. Aquatic Biodiversity in India: The Present Scenario. Daya Publishing House, New Delhi.

Khanna, D.R. and P.R. Yadav. Biology of Fishes. Discovery Publishing House, New Delhi.

Khanna, S.S. An Introduction of Fishes. Central Book Depot, Allahabad.

Khanna, S.S. and H.R. Singh. Text Book of Fish Biology and Fisheries. Narendra Publishing House, New Delhi.

Khuntia, B S. Postmortem Changes in Fish. Daya Publishing House, New Delhi.

Korakandy, R. Economics of Fisheries Management: A Critique in Third World Perspective. Daya Publishing House, New Delhi.

Korakandy, Ramakrishnan. Recreational Fisheries Development in India. Daya Publishing House, New Delhi.

Korakandy, Ramakrishnan. Technological Change and the Development of Marine Fishing Industry in India: A Case Study of Kerala. Daya Publishing House, New Delhi.

Kumar, Arvind & Harbhajan Kaur. Ecobiology of Aquatic Insects. Daya Publishing House, New Delhi.

Kumar, Arvind & Pushpalata Dubey eds. Fish Management and Aquatic Environment. Daya Publishing House, New Delhi.

Kumar, Arvind ed. Aquatic Environment and Toxicology. Daya Publishing House, New Delhi.

Kumar, Arvind ed. Ecology and Ethology of Aquatic Biota in 2 Vols. Daya Publishing House, New Delhi.

Kumar, Arvind ed. Ecology of Plankton. Daya Publishing House, New Delhi.

Kumar, Har Darshan. Sustainability and Management of Aquaculture and Fisheries. Daya Publishing House, New Delhi.

Kyle, Harry M. The Biology of Fishes: New Introduction by Dr Vijay Dev Singh. Daya Publishing House, New Delhi.

Lal, S.S. A Text Book of Practical Zoology, Vol. 3. Rastogi Publication, Meerut.

Laureti, Edmondo. Fish and Fishery Products/FAO. Daya Publishing House, New Delhi.

Malik, Davendra S & Kamal Singh Negi. Mahseer Fish Bionomics and Population: Barrage Impact on Fish Biology. Daya Publishing House, New Delhi.

Mishra, Smriti Ratna ed. Limnological Research in India. Daya Publishing House, New Delhi.

Mishra, Smriti Ratna. Management of Aquatic Habitats. Daya Publishing House, New Delhi.

Mitra, Abhijit *et al*. Introduction to Marine Plankton. Daya Publishing House, New Delhi.

Munro, Ian S R. The Marine and Fresh Water Fishes of Ceylon. Daya Publishing House, New Delhi.

Nagabhushanam, R., V.R. Awad and R. Saroji. Laboratory Exercises in Animal Physiology. COSIP-ULP (Biology) Publication.

Naskar, Kumudranjan. Manual of Indian Mangroves. Daya Publishing House, New Delhi.

Nath, Surendra ed. Recent Advances in Fish Ecology Limnology and Eco Conservation Vol 03. Daya Publishing House, New Delhi.

Nath, Surendra ed. Recent Advances in Fish Ecology Limnology and Eco Conservation Vol 04. Daya Publishing House, New Delhi.

New, Michael B & Somsuk Singholka. Freshwater Prawn Farming: A Manual for the Culture of Macrobrachium Rosenbergii/FAO. Daya Publishing House, New Delhi.

New, Michael B & Somsuk Singholka. Meetha Jal Jhinga Palan: Macrobranchium Rosenbergii Palan ki ek Niyamavali/FAO. Daya Publishing House, New Delhi.

New, Michael B *et al* eds. Farm Made Aquafeeds/FAO. Daya Publishing House, New Delhi.

Norman C. Fassett. A Manual of Aquatic Plants. Agrobios (India), Jodhpur.

Pandey, Nityanand & Davendra S Malik. Integrated Fish Farming. Daya Publishing House, New Delhi.

Pascoe, S. Bycatch Management and the Economics of Discarding/ FAO. Daya Publishing House, New Delhi.

Petr, T. Interactions Between Fish and Aquatic Macrophytes in Inland Waters: A Review/FAO. Daya Publishing House, New Delhi.

Petr, Tomi ed. Fisheries in Irrigation Systems of Arid Asia/FAO. Daya Publishing House, New Delhi.

Petr, Tomi ed. Inland Fishery Enhancements/FAO. Daya Publishing House, New Delhi.

Plaskitt, F J W. Microscopic Fresh Water Life. Daya Publishing House, New Delhi.

Project Report on Breeding of Carps with Ovaprim in India. Special Publication No. 4. Asian Fisheries Society, Indian Branch C/O College of Fisheries, Mangalore, India.

Rajagopalsamy, C B T & V Ramadhas. Nutrient Dynamics in Freshwater Fish Culture System. Daya Publishing House, New Delhi.

Rao, K.S. Practical Ecology. Anmol Publication Pvt. Ltd., New Delhi.

Rao, N Subba. Economics of Fisheries: A Case Study of Andhra Pradesh. Daya Publishing House, New Delhi.

Reddy, M.S. and K.R.S.S. Rao. A Text Book of Aquaculture. Discovery Publishing House, New Delhi.

Reddy, P V G K. Genetic Resources of Indian Major Carps/FAO. Daya Publishing House, New Delhi.

Roule, Louis. Fishes: Their Journeys and Migrations. Daya Publishing House, New Delhi.

Roule, Louis. Fishes: Their Ways of Life. Daya Publishing House, New Delhi.

Sakhare, Vishwas B ed. Advances in Aquatic Ecology Vol 1. Daya Publishing House, New Delhi.

Sakhare, Vishwas B ed. Advances in Aquatic Ecology Vol 2. Daya Publishing House, New Delhi.

Sakhare, Vishwas B. Applied Fisheries. Daya Publishing House, New Delhi.

Sakhare, Vishwas B. Ecology of Lakes and Reservoirs. Daya Publishing House, New Delhi.

Santhanam, R *et al*. A Manual of Freshwater Ecology: An Aspect of Fishery Environment. Daya Publishing House, New Delhi.

Santhanam, R. Fisheries Science. Daya Publishing House, New Delhi.

Sarkar, S K. Freshwater Fish Culture Vol 1. Daya Publishing House, New Delhi.

Sathiadhas, R. Production and Marketing Management of Marine Fisheries in India. Daya Publishing House, New Delhi.

Satyanarayana, D. Marine Chemistry. Daya Publishing House, New Delhi.

Saxena, Amita ed. Aquarium Management. Daya Publishing House, New Delhi.

Saxena, Amita ed. Control of Biotic and Abiotic Factors in Aquaculture. Daya Publishing House, New Delhi.

Secretan, P A D & C E Nash. Aquaculture and Risk Management: An FAO Fishing Manual/FAO. Daya Publishing House, New Delhi.

Seijo, J C *et al*. Fisheries Bioeconomics: Theory Modeling and Management/FAO. Daya Publishing House, New Delhi.

Seshappa, G. Indian Marine Biology. Daya Publishing House, New Delhi.

Sewell, R B Seymour. The Copepoda of Indian Seas. Daya Publishing House, New Delhi.

Shukla, G.S. and V.B. Upadhyay. Economic Zoology. Rastogi Publication, Meerut.

Singh, A K & Anil Mittal. Dictionary of Aquaculture. Daya Publishing House, New Delhi.

Singh, A K. Machali Palan Ke Naye Ayam. Daya Publishing House, New Delhi.

Singh, R K P. Economics of Aquaculture. Daya Publishing House, New Delhi.

Srivastava, C.B.L. A Text Book of Fishery Science and Indian Fisheries. Kitab Mahal, Allahabad.

Subasinghe, Rohana P *et al*. Health Management in Asian Aquaculture/FAO. Daya Publishing House, New Delhi.

Sugunan, V V. Reservoir Fisheries of India/FAO. Daya Publishing House, New Delhi.

Thomas, P C *et al*. Breeding and Seed Production of Fin Fish and Shell Fish. Daya Publishing House, New Delhi.

Unni, K Sankaran. Conservation and Management of Aquatic Ecosystems. Daya Publishing House, New Delhi.

Upadhyay, Alka Rani. Aquatic Plants for the Wastewater Treatment. Daya Publishing House, New Delhi.

Vijaykumar, K. Freshwater Ecosystem of India. Daya Publishing House, New Delhi.

Wanganeo, Ashwani & R K Langer eds. Trends in Biodiversity and Aquaculture. Daya Publishing House, New Delhi.

Yadav, B N. Fish and Fisheries 2nd Revised and Enlarged edn. Daya Publishing House, New Delhi.

Yadav, B N. Fish Endocrinology. Daya Publishing House, New Delhi.

Index